McCaulay's Practice Exams for the SAT* Subject Test in Mathematics Level 2

By Philip Martin McCaulay

Copyright Page

*SAT is a trademark registered by the College Board, which was not involved in the production of, and does not endorse, this product.

© Copyright 2012. All rights reserved. No part of this publication may be reproduced or transmitted in any form or by any means, electronic or mechanical, including photocopy, recording, or any information storage or retrieval system, without permission in written form from the publisher.

McCaulay's Practice Exams for the SAT* Subject Test in Mathematics Level 2
By Philip Martin McCaulay
Published in North Charleston, South Carolina, USA

Second Edition

Preface

McCaulay's Practice Exams for the SAT* Subject Test in Mathematics Level 2 contains two complete 50-question sample tests, for a total of 100 practice multiple-choice questions with answers and explanations. The Mathematics Level 2 test sponsored by the College Board has four categories of questions: Number and Operations; Algebra and Functions; Geometry and Measurement; and Data Analysis, Statistics, and Probability. The questions are arranged with the easier questions first and the more difficult questions at the end. The numbers of questions by topic for the practice tests in this book are shown in the following chart:

Topic	Total	Percent
Number and Operations	5 to 7	10% to 14%
Algebra and Functions	24 to 26	48% to 52%
Geometry and Measurement	14 to 16	28% to 32%
Data Analysis, Statistics, and Probability	4 to 6	8% to 12%
Total	50	100%

The Geometry and Measurement topic has three subtopics: Coordinate; Three-Dimensional; and Trigonometry. The numbers of questions by subtopic are shown in the following chart:

Topic	Total	Percent
Coordinate	5 to 7	10% to 14%
Three-Dimensional	2 to 3	4% to 6%
Trigonometry	6 to 8	12% to 16%
Total Geometry and Measurement	14 to 16	28% to 32%

Each multiple-choice question has five possible answers. For scoring, there is one point given for each correct answer to a question; no points for questions that are omitted; and minus one-fourth of a point for a wrong answer:

Answer	Points
Correct	1
Omit	0
Wrong	−0.25

About 55% to 65% of the questions require a calculator such as the Texas Instruments TI-83 Plus Graphing Calculator. On some questions, a graphing calculator may provide an advantage over a scientific calculator.

Content Overview

The following concepts are covered in the SAT* Subject Test in Mathematics Level 2:

Number and Operations (10% to 14%)
- Operations
- Ratio and Proportion
- Complex Numbers
- Counting
- Elementary Number Theory
- Matrices
- Sequences
- Series
- Vectors

Algebra and Functions (48% to 52%)
- Expressions
- Equations
- Inequalities
- Representation and Modeling
- Properties of Functions
    - Linear
    - Polynomial
    - Rational
    - Exponential
    - Logarithmic
    - Trigonometric
    - Inverse Trigonometric
    - Periodic
    - Piecewise
    - Recursive
    - Parametric

Geometry and Measurement (28% to 32%)
- Coordinate (10% to 14%)
  - Lines
  - Parabolas
  - Circles
  - Ellipses
  - Hyperbolas
  - Symmetry
  - Transformations
  - Polar Coordinates
- Three-Dimensional (4% to 6%)
  - Surface area and volume of solids
    - Cylinders
    - Cones
    - Pyramids
    - Spheres
    - Prisms
  - Coordinates in three dimensions
- Trigonometry (12% to 16%)
  - Right Triangles
  - Identities
  - Radian Measure
  - Law of Cosines
  - Law of Sines
  - Equations
  - Double Angle Formulas

Data Analysis, Statistics and Probability (8% to 12%)
- Mean, Median, Mode, and Range
- Interquartile Range
- Standard Deviation
- Graphs and Plots
- Least-Squares Regression
  - Linear
  - Quadratic
  - Exponential
- Probability

Table of Contents

Math Level 2 Exam 1 1

Math Level 2 Exam 1 Answers 31

Math Level 2 Exam 1 Explanations 32

Math Level 2 Exam 2 94

Math Level 2 Exam 2 Answers 124

Math Level 2 Exam 2 Explanations 125

About the Author 186

Math Level 2 Exam 1

McCaulay's Practice Exams for the SAT* Subject Test in Mathematics Level 2

Math Level 2 Exam 1
Time – 1 Hour
50 Questions

Directions: For this section, solve each problem and decide which answer is the best. Numerical answers may be rounded. A scientific or graphing calculator will be necessary for some questions. Angles are in degrees. Unless specified otherwise, the domain of any function is the set of all real numbers. Figures are not necessarily drawn to scale.

Reference Information

Volume of a right circular cone with radius r and height h: $V = \frac{1}{3} \pi r^2 h$

Volume of a sphere with radius r: $V = \frac{4}{3} \pi r^3$

Volume of a pyramid with base area B and height h: $V = \frac{1}{3} B h$

Surface area of a sphere with radius r: $S = 4 \pi r^2$

---

1. The height of a right circular cylinder is equal to 5 times the diameter. If the volume of the cylinder is 25, what is the height of the cylinder?

    (A)   4.63

    (B)   6.14

    (C)   9.27

    (D)   13.90

    (E)   18.53

2. If $f(x) = |3 - 2x|$, then $f(4) =$

   (A) $f(-3/2)$

   (B) $f(-1)$

   (C) $f(-1/2)$

   (D) $f(2/3)$

   (E) $f(3/2)$

3. What is the distance in space between the points with coordinates $(-4, 11, -3)$ and $(7, 9, -5)$?

   (A) 30

   (B) 35

   (C) 8.77

   (D) 11.36

   (E) 17.35

4. A game has a pair of dice. One die is a traditional cube with six sides numbered 1 through 6, inclusive, with each value being equally likely. The second die is a 24-sided polyhedron numbered 1 through 24, inclusive, with each value being equally likely. What is the probability that the six-sided die rolls a six and the 24-sided die rolls a number that is not a six?

   (A) $1/144$

   (B) $23/144$

   (C) $1/6$

   (D) $24/138$

   (E) $25/144$

5.

Which of the following functions could describe the above sequence, given that the first term is defined as $f(0) = 1$?

(A) $f(x) = (x)(f(x-1))$

(B) $f(x) = (x-1)(f(x-1))$

(C) $f(x) = (x+1)(f(x-1))$

(D) $f(x) = (2)(f(x-1)) + (x-1)$

(E) $f(x) = (2x)(f(x-1)) - 2$

6. The height and diameter of a right circular cone are equal. If the volume of the cone is 7, what is the height of the cone?

(A) 2.7

(B) 3.0

(C) 3.3

(D) 3.7

(E) 4.3

7. Evaluate $sin\,(arccos\,0.76)$ in radians.

(A) 0.57

(B) 0.61

(C) 0.64

(D) 0.65

(E) 0.69

8. In the xy-plane, what is the area of a triangle whose vertices are ($\sqrt{3}$, 0), ($\sqrt{7}$, −4), and ($\sqrt{5}$, −4) ?

   (A)   0.8

   (B)   1.6

   (C)   2.5

   (D)   3.2

   (E)   5.0

9. Currently there are four workers for each Medicare beneficiary. The number of Medicare beneficiaries is expected to grow exponentially at a rate of 1.4% per year. The number of workers is expected to remain the same. Fifty years from now, how many workers will there be for each Medicare beneficiary?

   (A)   1

   (B)   2

   (C)   3

   (D)   4

   (E)   5

10. How many ways can two teams be made of 13 players, one team with 6 players and the other team with 7 players?

    (A)   610

    (B)   987

    (C)   1,716

    (D)   2,584

    (E)   4,181

11. If $f(x) = \dfrac{5 - 4x}{7x + 11}$, what value does $f(x)$ approach as x gets infinitely larger?

    (A)   −0.571
    (B)   −0.056
    (C)   0.000
    (D)   0.065
    (E)   0.455

12.

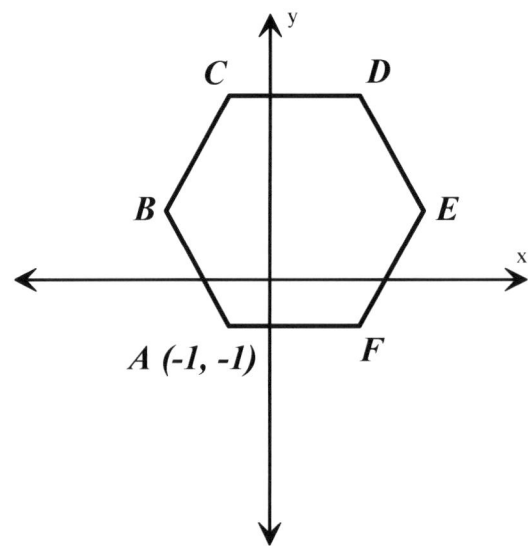

In the figure above, *ABCDEF* is a regular hexagon with sides of length 3. What is the *y*-coordinate of *E* ?

    (A)   1.1
    (B)   1.6
    (C)   1.8
    (D)   2.5
    (E)   3.5

13.

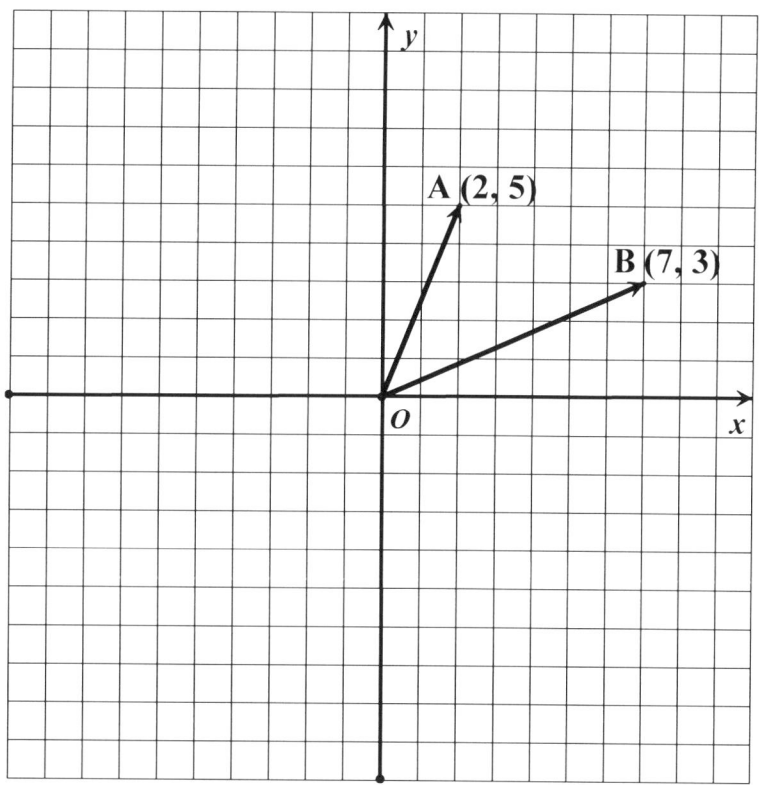

In the figure above, when $\vec{OA}$ is subtracted from $\vec{OB}$, what is the length of the resultant vector?

(A) 2.2

(B) 4.0

(C) 5.4

(D) 5.8

(E) 7.6

14. The number tau, τ, is defined as two times pi, π. What is the degree measure of an angle whose radian measure is $\dfrac{5\tau}{8}$?

   (A)   56.25°

   (B)   112.5°

   (C)   135°

   (D)   225°

   (E)   270°

15. If $\cos\theta = 0.974$, then $\cos(\pi + \theta) =$

   (A)   −0.974

   (B)   −0.228

   (C)   0.228

   (D)   0.570

   (E)   0.974

16. Evaluate *arcsin (sin (π / 2))* in radians.

   (A)   0

   (B)   1

   (C) $\pi / 2$

   (D) π

   (E) 2π

17. The average price of the cars on a lot is $37,180. The median is $35,000 and the standard deviation is $7,000. The dealer reduces the price of each car by $1,000. Which of the following statements are true?

   I.   The new mean is $36,180.

   II. The median remains $35,000.

   III The standard deviation remains $7,000.

   (A)   I only

   (B)   I and II only

   (C)   II and III only

   (D)   I and III only

   (E)   I, II, and III

18. Which of the following graphs represents the function

   $f(x) = 0.4 \cos(3x - 2\pi) + 1$?

   (A)

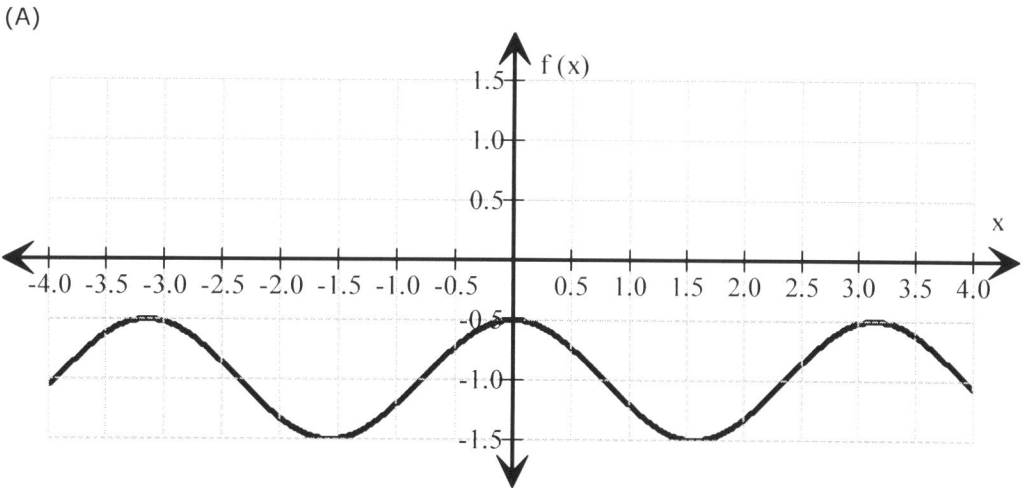

(B)

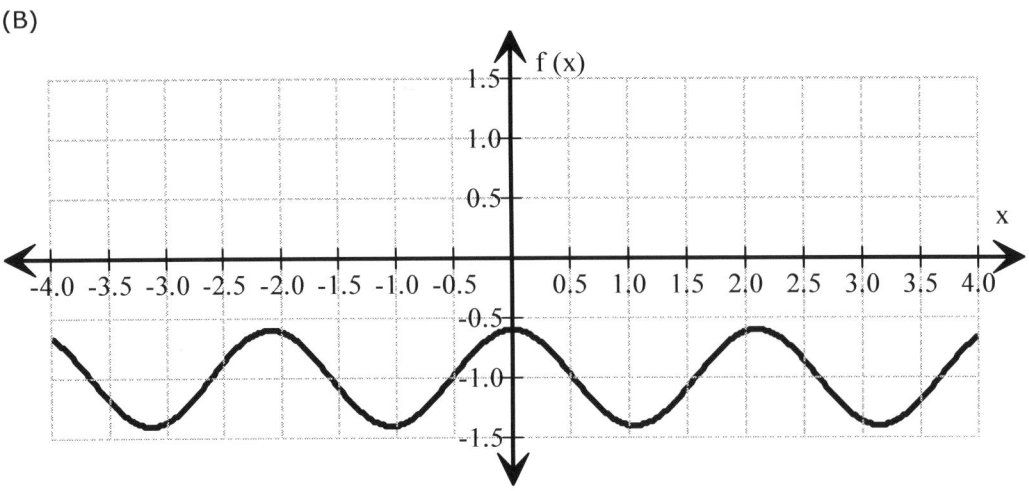

(C)

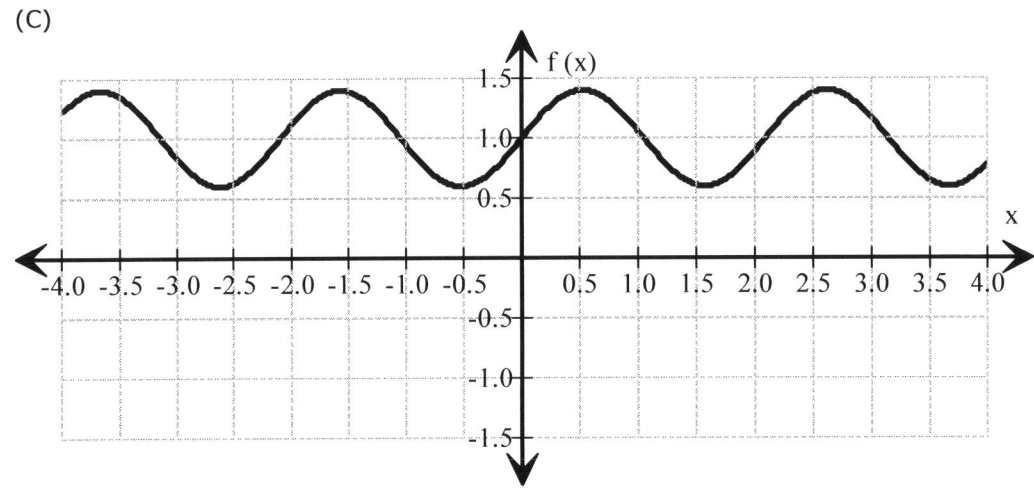

(D)

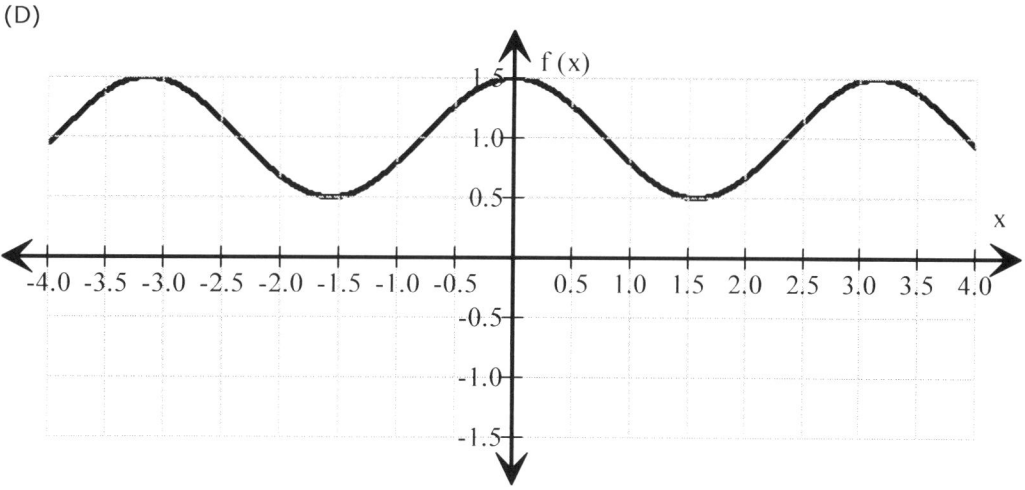

(E)

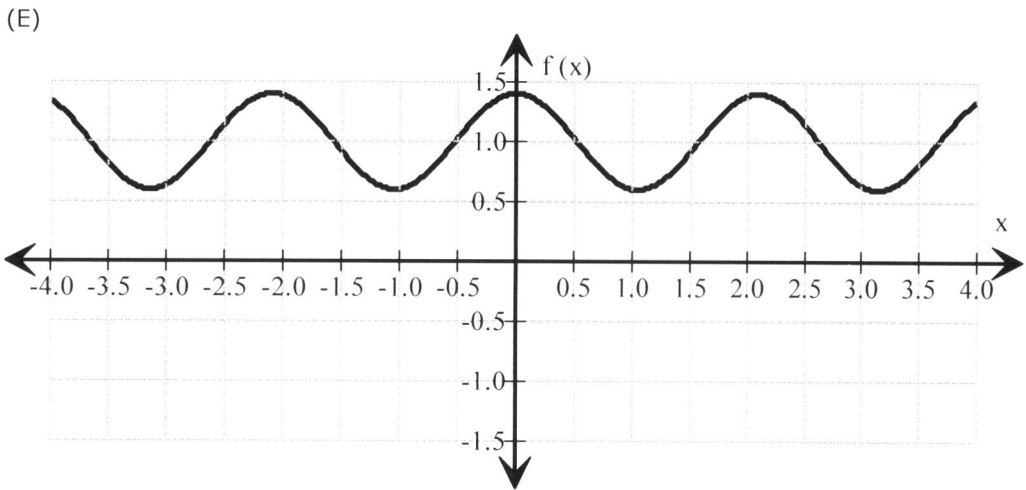

19. In a group of 30 people, 10 percent are left-handed, and the rest are right-handed. Three people are to be selected at random from the group. What is the probability that all three will be right-handed?

(A)   0.72

(B)   0.73

(C)   0.81

(D)   0.89

(E)   0.90

20. The formula $N = N_0 / \ln(4t)$ gives the population of a decaying colony of bacteria after t hours for an initial population of N_0, where t is at least equal to 1. How many hours will it take for the population to decrease by 70%?

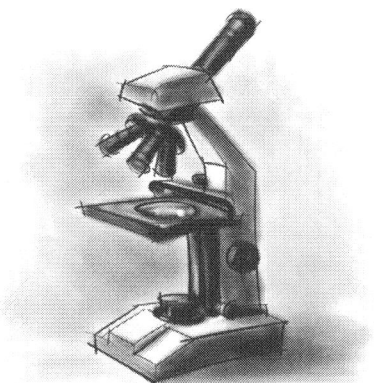

(A)   1

(B)   2

(C)   4

(D)   6

(E)   7

21. For some positive integer t, the first three terms of a geometric sequence are $5t + 1$, $1 - 3t$, and $2t - 2$. What is the numerical value of the fourth term?

(A)   −4

(B)   −2

(C)   0

(D)   2

(E)   4

22. An initial investment of $1,000 increases exponentially to $1,500 after 5 years. At the same rate, in how many years will the value be at least $4,500?

   (A)   15
   (B)   17
   (C)   18
   (D)   19
   (E)   20

23. If $0 \leq \theta < \pi/2$ and $\sin \theta = 3 \cos \theta$, what is the value of θ ?

   (A)   0.24
   (B)   0.32
   (C)   1.11
   (D)   1.25
   (E)   1.33

24. What is the range of the function defined by $f(x) = \dfrac{1}{(x-2)(x+1)}$?

   (A)   All real numbers.
   (B)   All real numbers except -1 and 2.
   (C)   All real numbers greater than 0 or less than or equal to $-4/9$.
   (D)   All real numbers greater than 0 or less than or equal to -0.5.
   (E)   All real numbers greater than 0 or less than -2.25.

25. What is the sum of the infinite geometric series $6 + 2 + \frac{2}{3} + \frac{2}{9} + \ldots$?

(A)　9

(B)　12

(C)　13.5

(D)　15

(E)　18

26. Which of the following graphs could be the parametric representation of the equations $x = a\ \cos(t)$ and $y = a\ \sin(t)$ where $0 \leq t \leq 2$?

(A)

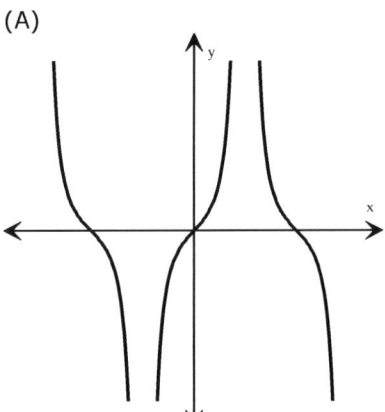

(B)

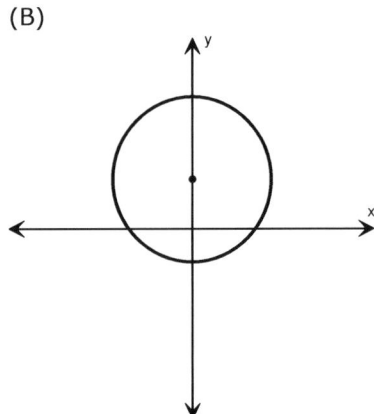

(C)

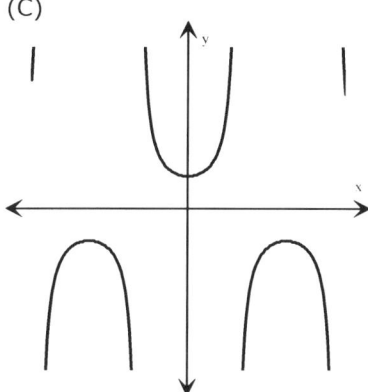

(D)

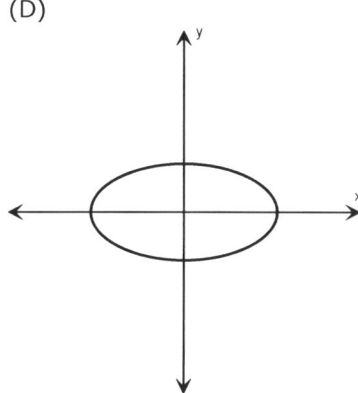

(E)

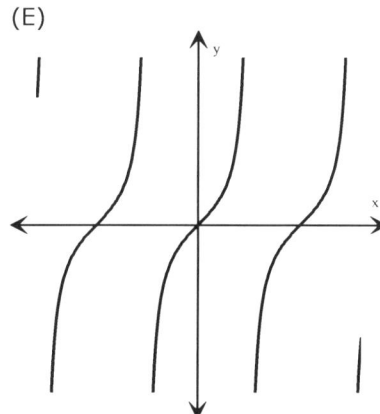

27.

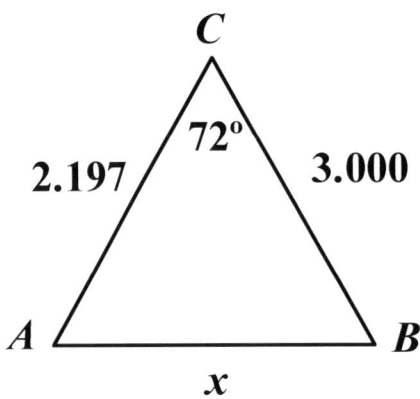

Note: Figure not drawn to scale.

In the figure above, AC = 2.197, BC = 3.000, and ∠ C = 72°. What is the value of x ?

(A) 2.53

(B) 2.83

(C) 3.08

(D) 3.12

(E) 3.28

28. The Fibonacci sequence is recursively defined by $a_n = a_{n-1} + a_{n-2}$, for $n > 2$. If $a_1 = 1$ and $a_2 = 1$, what is the value of a_{17} ?

(A) 610

(B) 987

(C) 1,597

(D) 2,584

(E) 4,181

29.
800, 300, 770, 710, 700, 650, 650, 600, 560, 760

A group of students had the above scores on a college entrance exam. If the percentiles are in the range 0 to 100, inclusive, what is the interquartile range for the scores?

(A)   60
(B)   135
(C)   160
(D)   173
(E)   185

30. What annual compounded rate of return would result in an initial investment of $1,000 accumulating to $4,000 in 18 years?

(A)   7.2%
(B)   7.3%
(C)   7.7%
(D)   8.0%
(E)   8.6%

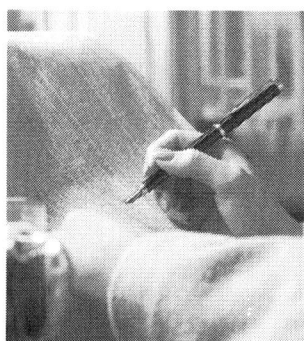

31. Find the exponential function that best fits the data below.

x	0.1	1.2	1.9	3.1	4.2	4.9
y	0.60	0.70	0.90	1.15	1.42	1.82

(A) $y = (0.57)(1.24^x)$
(B) $y = (0.57)(1.25^x)$
(C) $y = (0.57)(1.26^x)$
(D) $y = (0.58)(1.25^x)$
(E) $y = (0.58)(1.26^x)$

32. If $f(x) = \sqrt[4]{\dfrac{x^2}{3} + 2}$, what is $f^{-1}(1.8)$?

(A) 1.05

(B) 1.79

(C) 2.91

(D) 4.70

(E) 5.05

33. Using the quadratic regression model, which parabola best fits the data for the number of new homes sold in a community?

Year	0	1	2	3	4	5	6
Number Sold	98	76	58	54	60	74	99

(A) $y = -4.8x^2 + 29.5x - 98.6$

(B) $y = -4.7x^2 + 29.4x - 98.5$

(C) $y = 4.7x^2 - 29.4x + 98.5$

(D) $y = 4.8x^2 - 29.5x + 98.6$

(E) $y = 4.9x^2 - 29.6x + 98.7$

34. What value does $\dfrac{2\ln(x)}{1-x}$ approach as x approaches 1?

(A) -2

(B) -1

(C) 0

(D) 1

(E) 2

35. Which of the following statements are accurate regarding symmetry of functions?

   I. There is symmetry about the x-axis if $f(x) = f(-x)$

   II. There is symmetry about the y-axis if $f(x) = -f(x)$

   III. There is symmetry about the origin if $f(x) = -f(-x)$

(A) I only

(B) II only

(C) III only

(D) I and II only

(E) I and III only

36. The function f is defined by $f(x) = \dfrac{x^3}{3} - \dfrac{3x^2}{4} - \dfrac{x}{2} + 2$ for $-1 \leq x \leq 2$. What is the difference between the maximum and minimum values of (x) ?

(A) 1.33

(B) 1.41

(C) 1.42

(D) 1.46

(E) 1.50

37.

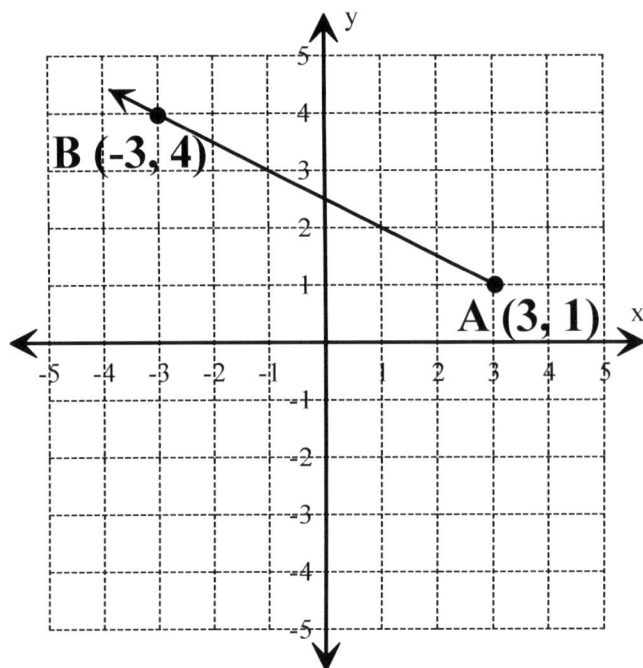

Which of the following vector equations is represented by the above graph, given that an object moves from point A to point B in 2 seconds?

(A) $y = (3, 1) + t(-3, 1.5)$

(B) $y = (3, 1) + t(-2, 1)$

(C) $y = (3, 1) - t(1, -0.5)$

(D) $y = (5, 0) + t(-4, 2)$

(E) $y = (5, 0) - t(4, 2)$

38. What is the period of the mantissa function, $f(x) = x - [x]$, where $[x]$ is the greatest integer less than or equal to x ?

(A) -1

(B)   0

(C)   1

(D) $\sqrt{2}$

(E)   2

39.

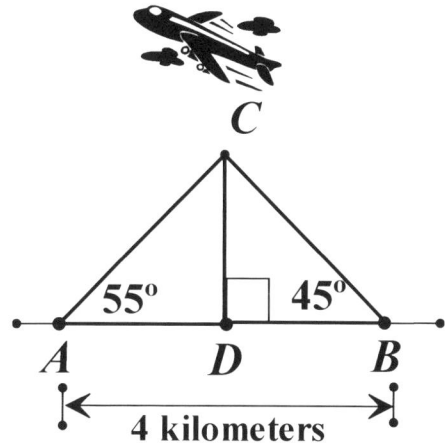

Note: Figure not drawn to scale.

The airplane in the figure above is directly over point D at point C on a level runway. The angles of elevation for points A and B are 55° and 45°, respectively. If points A and B are 4 kilometers apart, what is the distance, in kilometers, from point B to the airplane?

(A)   1.65

(B)   2.35

(C)   2.83

(D)   2.87

(E)   3.33

40. Which of the following graphs is the best parametric representation of the equations $x = -5\cos(t)$ and $= \sin(t)$?

(A)

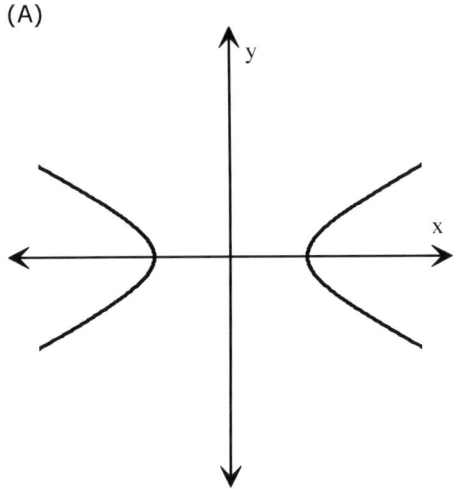

(B)

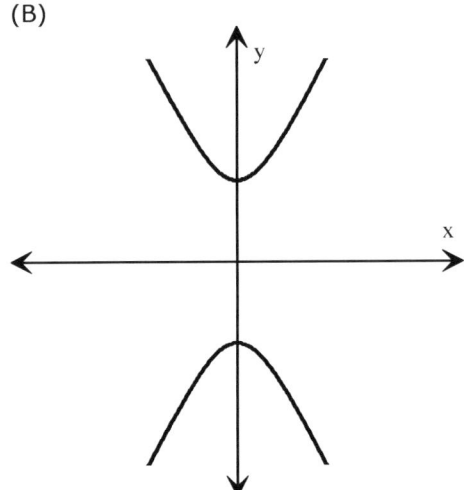

(C)

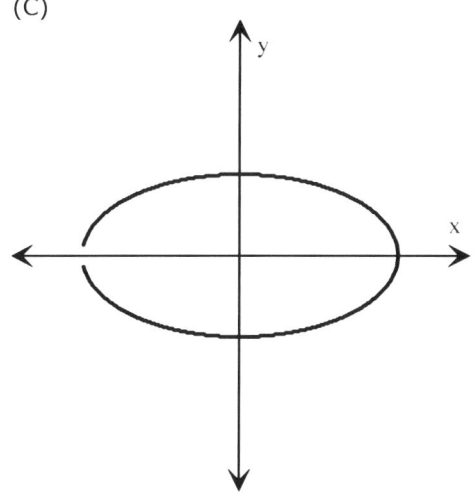

(D)

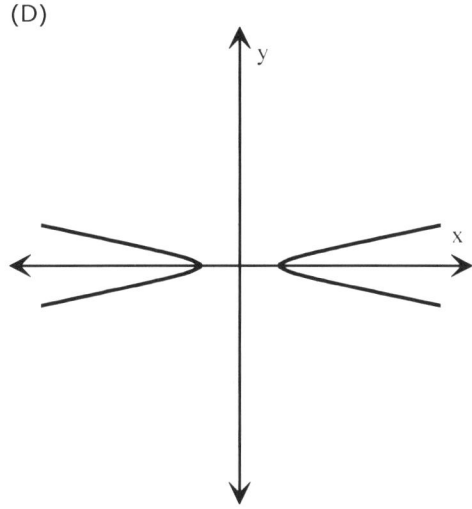

(E)

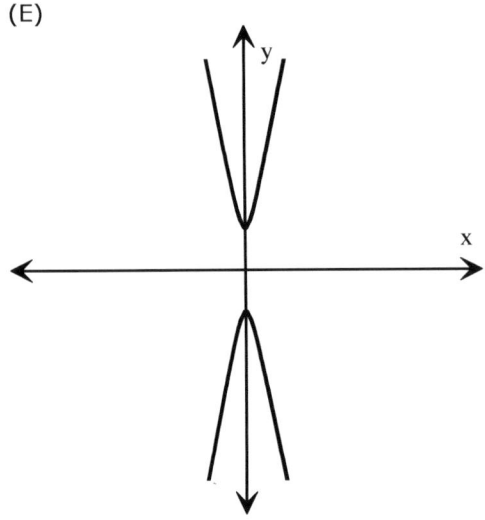

41. Which of the following expressions is equivalent to $\sin(\pi + 2\theta)$?

    (A) $\sin^2 \theta$

    (B) $\sin \theta \cos \theta$

    (C) $-\sin \theta \cos \theta$

    (D) $2 \sin \theta \cos \theta$

    (E) $-2 \sin \theta \cos \theta$

42.
$$f(x) = 0.5 \sin(nx + \pi) + 1.25$$

For what value of n, where $n > 0$, is the period of the above function equal to 4π?

    (A) 0.25

    (B) 0.50

    (C) 0.75

    (D) 1.00

    (E) 1.25

43. Suppose the graph of $f(x) = 1 - x^2$ is translated 2 units right and 3 units down. If the resulting graph is $g(x)$, what is the value of $g(-2)$?

   (A) -18
   (B) -11
   (C) -5
   (D) -2
   (E) 3

44.

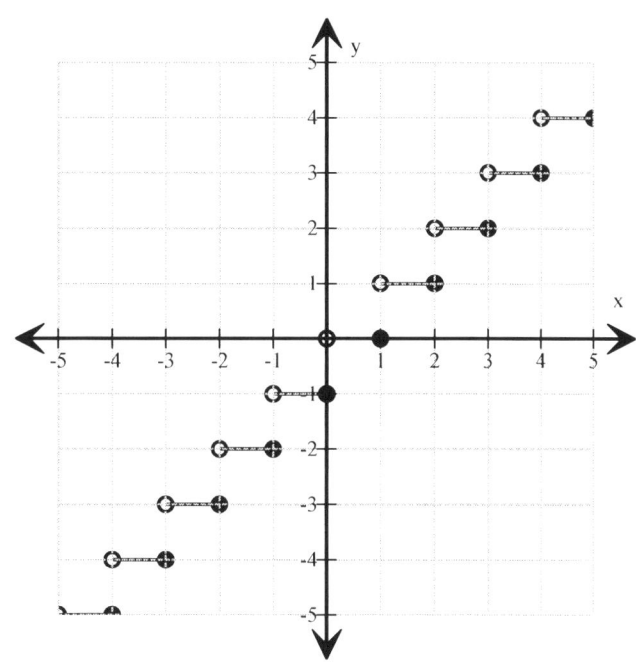

Which of the following expressions best describes the above graph?

   (A)   The floor function or greatest integer function $y = [x]$ where y is the greatest integer not less than x

   (B)   The ceiling function where y is the smallest integer not less than x

   (C)   The function where y is the smallest integer not less than $x - 1$

   (D)   The function where y is the greatest integer not less than $x - 1$

   (E)   The mantissa function $y = x - [x]$

45. Which of the following graphs could be the parametric representation of the equations $x = 2\cos(t)$ and $y = 4\sin(t)$?

(A)

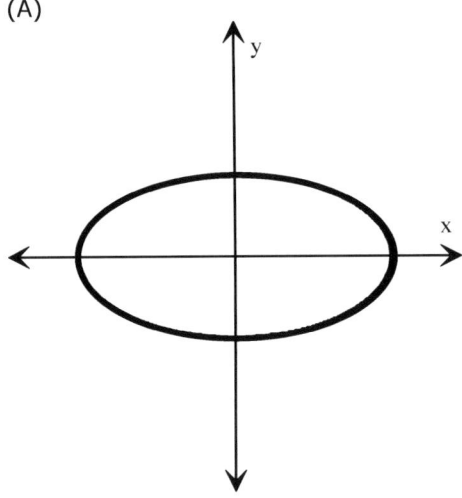

(B)

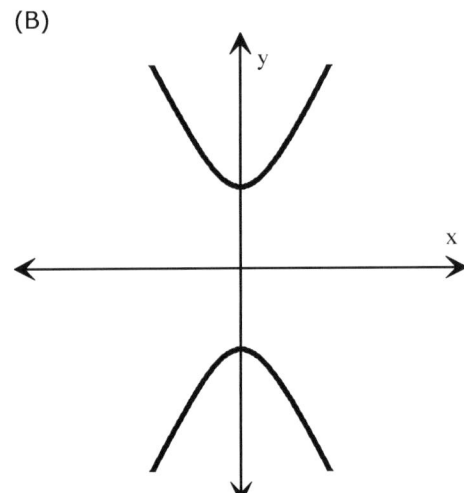

(C)

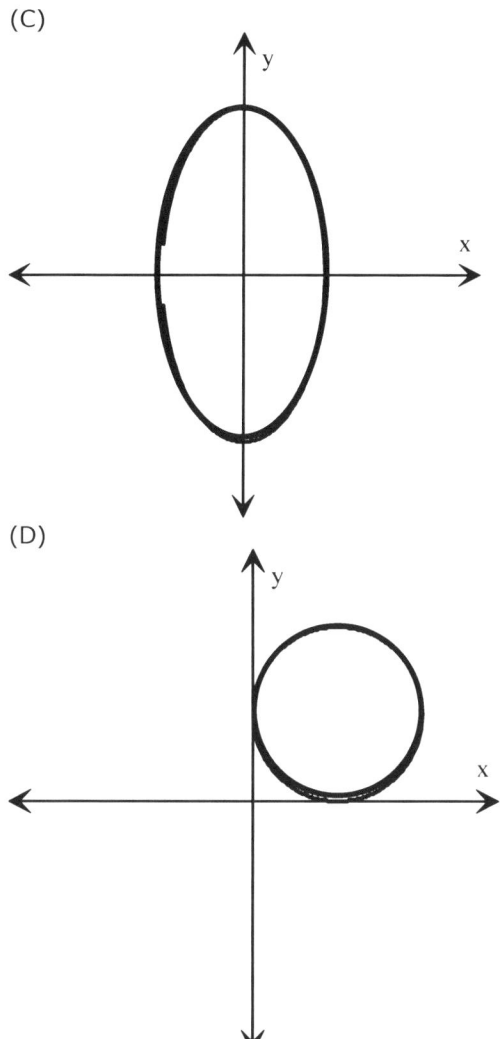

(D)

(E)

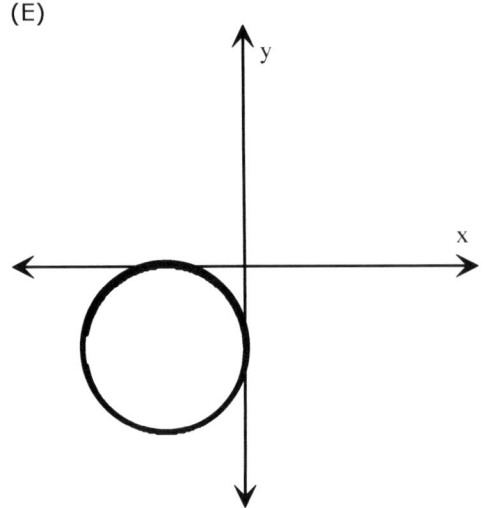

46. Express the point with coordinates (2.76, 2.31) in terms of polar coordinates.

(A) (1.5, 40°)

(B) (1.5, 50°)

(C) (3.6, 30°)

(D) (3.6, 40°)

(E) (3.6, 50°)

47. Which of the following statements are true regarding inverse trigonometric functions?

I. The domain of $y = sin^{-1} x$ is $-1 \leq x \leq 1$, and the range is $-\frac{\pi}{2} \leq y \leq \frac{\pi}{2}$.

II. The domain of $y = cos^{-1} x$ is $-1 \leq x \leq 1$, and the range is $0 \leq y \leq \pi$.

III The domain of $y = tan^{-1} x$ is all real numbers, and the range is $0 \leq y \leq \pi$.

(A) I only

(B) I and II only

(C) II and III only

(D) I and III only

(E) I, II, and III

48. A bond pays annual coupons of $600 at the end of each year for 10 years and returns the original $10,000 investment at the end of the 10 years. What is the present value of the bond, with the cash flows discounted using an interest rate of 3%?

(A) $12,342

(B) $12,496

(C) $12,559

(D) $12,713

(E) $12,871

49.

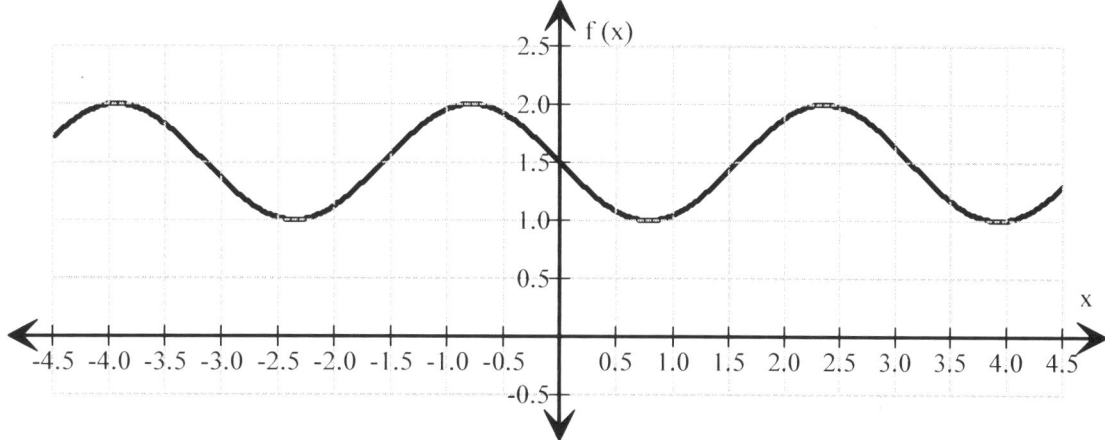

The graph above could represent which of the following functions?

(A) $f(x) = 2 \sin(0.5x + 2\pi) - 1.5$

(B) $f(x) = 2 \sin(0.5x + \pi) + 1.5$

(C) $f(x) = 2 \sin(0.5x + 2\pi) + 1.5$

(D) $f(x) = 0.5 \sin(2x + \pi) - 1.5$

(E) $f(x) = 0.5 \sin(2x + \pi) + 1.5$

50. What is the equation of the hyperbola having foci at (1, 8.4) and (1, −4.4) and vertices (1, 7) and (1, −3)?

   (A) $(1/5)(x-2)^2 - (1/4)(y-1)^2 = 1$

   (B) $(1/5)(y-2)^2 - (1/16)(x-1)^2 = 1$

   (C) $(1/25)(y-2)^2 - (1/4)(x-1)^2 = 1$

   (D) $(1/25)(y-2)^2 - (1/16)(x-1)^2 = 1$

   (E) $(1/25)(y-2)^2 + (1/16)(x-1)^2 = 1$

Math Level 2 Exam 1 Answers

1	C	21	B	41	E
2	B	22	D	42	B
3	D	23	D	43	A
4	B	24	C	44	C
5	A	25	A	45	C
6	B	26	B	46	D
7	D	27	D	47	B
8	A	28	C	48	C
9	B	29	B	49	E
10	C	30	D	50	C
11	A	31	C		
12	B	32	E		
13	C	33	E		
14	D	34	A		
15	A	35	C		
16	C	36	D		
17	D	37	A		
18	E	38	C		
19	A	39	E		
20	E	40	E̶ C		

McCaulay's Practice Exams for the SAT* Subject Test in Mathematics Level 2

Math Level 2 Exam 1 Explanations

1. The height of a right circular cylinder is equal to 5 times the diameter. If the volume of the cylinder is 25, what is the height of the cylinder?

    (A) 4.63

    (B) 6.14

    *(C) 9.27

    $V = \pi r^2 h$

    $h = 5d = 10r$

    $V = 25$

    $25 = \pi r^2 (10r) = 10 \pi r^3$

    $\pi r^3 = \dfrac{25}{10} = 2.5$

    $r = (2.5/\pi)^{1/3} \approx 0.927$

    $h = 10r = (10)(0.927) \approx 9.27$

    Geometry and Measurement; Surface Area and Volume of Solids, Cylinders

    (D) 13.90

    (E) 18.53

Page 32

2. If $f(x) = |3 - 2x|$, then $f(4) =$

   (A) $f(-3/2)$

   *(B) $f(-1)$

   $f(4) = |3 - 2(4)| = |3 - 8| = |-5| = 5$

   $f(n) = 3 - 2n = 5$

   $2n = -2$

   $n = -1$

   $f(4) = f(-1)$

   Algebra and Functions; Properties of Functions, Linear

   (C) $f(-1/2)$

   (D) $f(2/3)$

   (E) $f(3/2)$

3. What is the distance in space between the points with coordinates $(-4, 11, -3)$ and $(7, 9, -5)$?

(A) 30

(B) 35

(C) 8.77

*(D) 11.36

In three-dimensional Euclidean space, the distance between two points is

$$d(p,q) = \sqrt{(p_1 - q_1)^2 + (p_2 - q_2)^2 + (p_3 - q_3)^2}$$

$$d(p,q) = \sqrt{(-4 - 7)^2 + (11 - 9)^2 + (-3 - (-5))^2}$$

$$= \sqrt{(-11)^2 + (2)^2 + (2)^2} = \sqrt{121 + 4 + 4} = \sqrt{129} \approx 11.36$$

Geometry and Measurement; Coordinates in Three Dimensions

(E) 17.35

4. A game has a pair of dice. One die is a traditional cube with six sides numbered 1 through 6, inclusive, with each value being equally likely. The second die is a 24-sided polyhedron numbered 1 through 24, inclusive, with each value being equally likely. What is the probability that the six-sided die rolls a six and the 24-sided die rolls a number that is not a six?

(A) $1/144$

*(B) $23/144$

The probability of a six on the six-sided die = $1/6$

The probability of a six on the 24-sided die = $1/24$

The probability of a number that is not six on the 24-sided die

$= 1 - 1/24 = 23/24$

The probability that the six-sided die rolls a six and the 24-sided die rolls a number that is not six

$= (1/6)(23/24) = (23/144)$

Data Analysis, Statistics, and Probability; Probability

(C) $1/6$

(D) $24/138$

(E) $25/144$

5. 1, 1, 2, 6, 24, 120, ...

Which of the following functions could describe the above sequence, given that the first term is defined as $(0) = 1$?

*(A) $f(x) = (x)(f(x-1))$

x	x − 1	f(x − 1)	(x)(f(x − 1))
1	0	1	1
2	1	1	2
3	2	2	6
4	3	6	24
5	4	24	120

This is the recursive definition of the factorial function.

Algebra and Functions; Properties of Functions, Recursive

(B) $f(x) = (x-1)(f(x-1))$

(C) $f(x) = (x+1)(f(x-1))$

(D) $f(x) = (2)(f(x-1)) + (x-1)$

(E) $f(x) = (2x)(f(x-1)) - 2$

6. The height and diameter of a right circular cone are equal. If the volume of the cone is 7, what is the height of the cone?

(A)   2.7

*(B)   3.0

Volume of a right circular cone with radius r and height h: $V = \frac{1}{3}\pi r^2 h$

$h = d = 2r$

$V = 7$

$7 = \frac{1}{3}\pi r^2 (2r) = \frac{2}{3}\pi r^3$

$\pi r^3 = \frac{3}{2}(7) = 10.5$

$r = (10.5/\pi)^{1/3} = 1.495$

$h = 2r = (2)(1.495) \approx 3.0$

Geometry and Measurement; Surface Area and Volume of Solids, Cones

(C)   3.3

(D)   3.7

(E)   4.3

7. Evaluate *sin* (*arccos* 0.76) in radians.

    (A)   0.57

    (B)   0.61

    (C)   0.64

    *(D) 0.65

    $\cos \theta = 0.76$ when $\theta = 0.70748$

    $\sin (0.70748) \approx 0.65$

    Algebra and Functions; Properties of Functions, Inverse Trigonometric

    (E)   0.69

8. In the xy-plane, what is the area of a triangle whose vertices are $(\sqrt{3}, 0)$, $(\sqrt{7}, -4)$, and $(\sqrt{5}, -4)$?

*(A)   0.8

$$A = \frac{1}{2} b h$$

$$b = \sqrt{7} - \sqrt{5}$$

$$h = 0 - (-4) = 4$$

$$A = \frac{1}{2} \left(\sqrt{7} - \sqrt{5} \right) (4) = (2)\left(\sqrt{7} - \sqrt{5} \right) \approx (2)(0.41) \approx 0.8$$

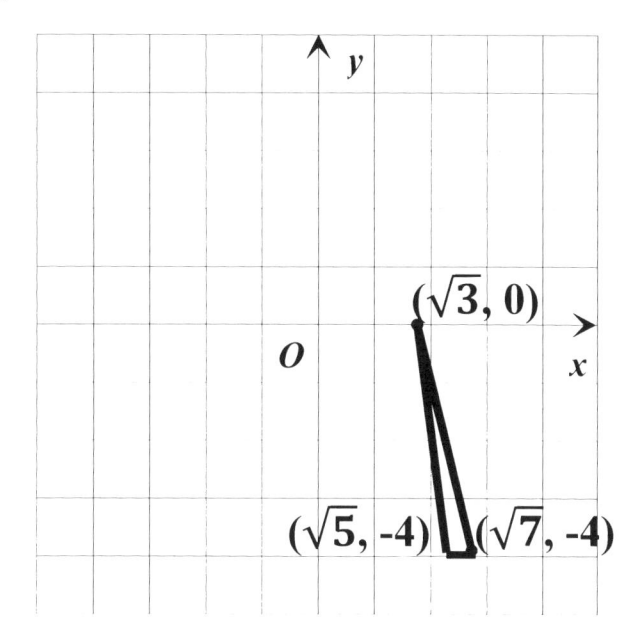

Geometry and Measurement, Coordinate, Lines

(B)   1.6

(C)   2.5

(D)   3.2

(E)   5.0

9. Currently there are four workers for each Medicare beneficiary. The number of Medicare beneficiaries is expected to grow exponentially at a rate of 1.4% per year. The number of workers is expected to remain the same. Fifty years from now, how many workers will there be for each Medicare beneficiary?

(A)   1

*(B)   2

(C)   3

(D)   4

There are 4 workers for each Medicare beneficiary.

$m / w = 1 / 4$

The number of Medicare beneficiaries grows by 1.4% compounded each year.

$(1.014)^{50} \approx 2$

The number of Medicare beneficiaries will double while the number of workers remains the same.

Algebra and Functions; Properties of Functions, Exponential

(E)   5

10. How many ways can two teams be made of 13 players, one team with 6 players and the other team with 7 players?

(A) 610

(B) 987

*(C) 1,716

A combination is a way of selecting several things from a group where order does not matter.

$$\binom{n}{k} = \frac{(n)(n-1)\ldots(n-k+1)}{(k)(k-1)\ldots(1)} = \frac{n!}{k!(n-k)!}$$

The factorial for a non-negative number *n* is denoted by *n* ! and is defined as the product of all positive integers less than or equal to *n*.

$$n! = (n)(n-1)\ldots(1)$$

0 ! = 1 1 ! = 1; 2! = 2; 3! = 6; 4! = 24; 5! = 120

$$\binom{13}{7} = \binom{13}{6} = \frac{13!}{7!\,6!}$$

$$= \frac{(13)(12)(11)(10)(9)(8)\,(7!)}{(7!)\,(6)(5)(4)(3)(2)}$$

$$= \frac{(13)(12)(11)(10)(9)(8)}{(6)(5)(4)(3)(2)}$$

$$= \frac{(13)(2)(11)(2)(9)}{(3)} = 1{,}716$$

Number and Operations; Counting

(D) 2,584

(E) 4,181

11. If $f(x) = \dfrac{5-4x}{7x+11}$, what value does $f(x)$ approach as x gets infinitely larger?

*(A)   −0.571

$$f(1{,}000{,}000) = \dfrac{5 - 4(1{,}000{,}000)}{7(1{,}000{,}000) + 11} = \dfrac{-3{,}999{,}995}{7{,}000{,}011} \approx -0.571$$

$$\lim_{n \to \infty}\left(\dfrac{5-4x}{7x+11}\right) = -\dfrac{4}{7} \approx -0.571$$

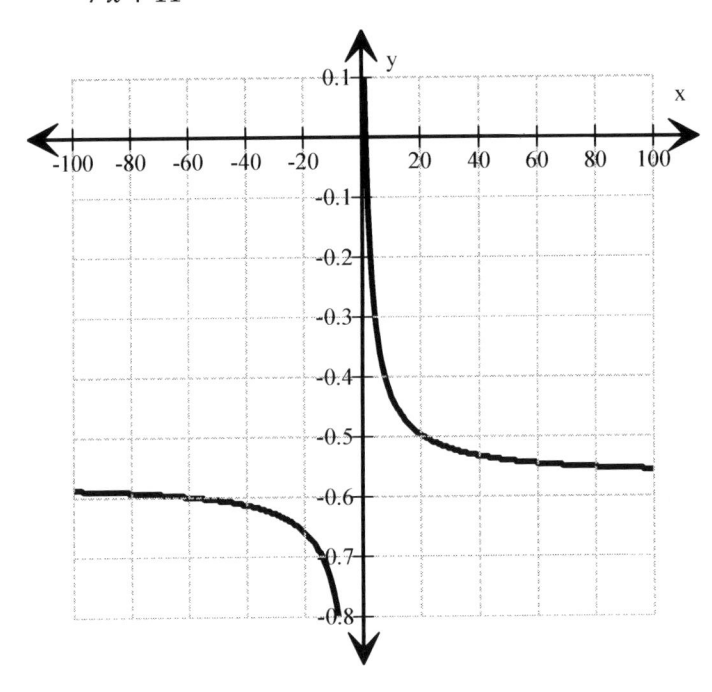

Algebra and Functions; Properties of Functions, Rational

(B)   −0.056

(C)   0.000

(D)   0.065

(E)   0.455

12.

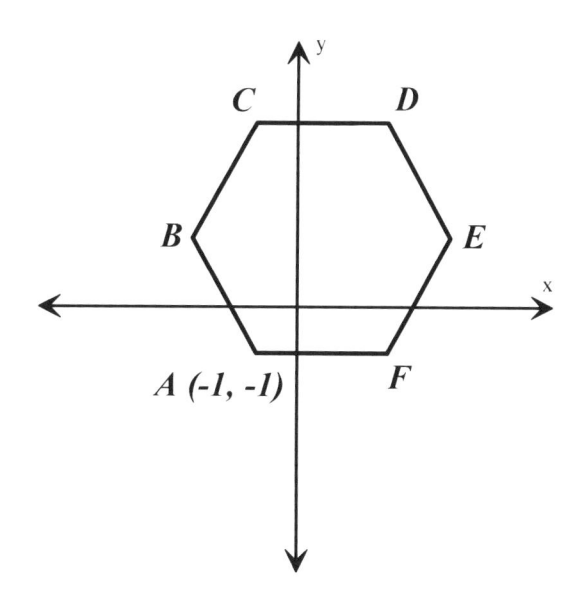

In the figure above, *ABCDEF* is a regular hexagon with sides of length 3. What is the *y*-coordinate of *E* ?

(A)   1.1

*(B)   1.6

The sum of the interior angles of a hexagon = (6 − 2) (180°) = 720°.

Each interior angle measures 720° / 6 = 120°.

Angle *AFD* = 90°.

Angle *DFE* = 120° − 90° = 30°.

Angle *DEB* = Angle *BEF* = 120° / 2 = 60°

EF is the hypotenuse of a 30°: 60°: 90° triangle with sides in the ratio 1: √3 : 2. *EF* has length 3.

The sides of the triangle are 1.5, 1.5 √3, and 3.

The coordinates of *E* are (−1 + 3 + 1.5, −1 + 1.5 √3) ≈ (3.5, 1.6)

The *y*-coordinate = 1.6

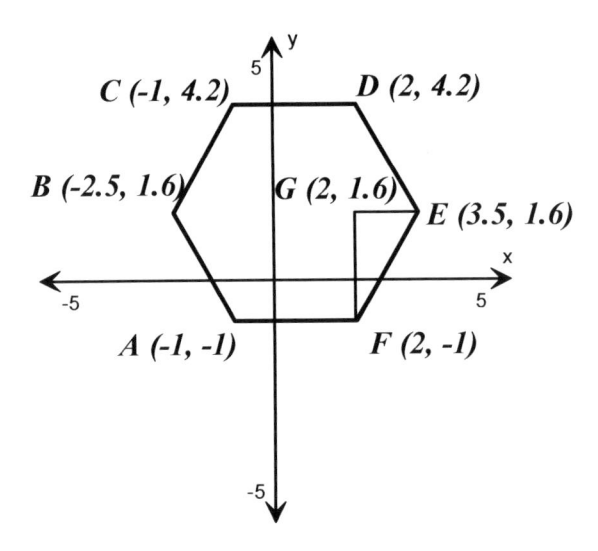

Geometry and Measurement; Coordinate, Lines

(C)   1.8

(D)   2.5

(E)   3.5

13.

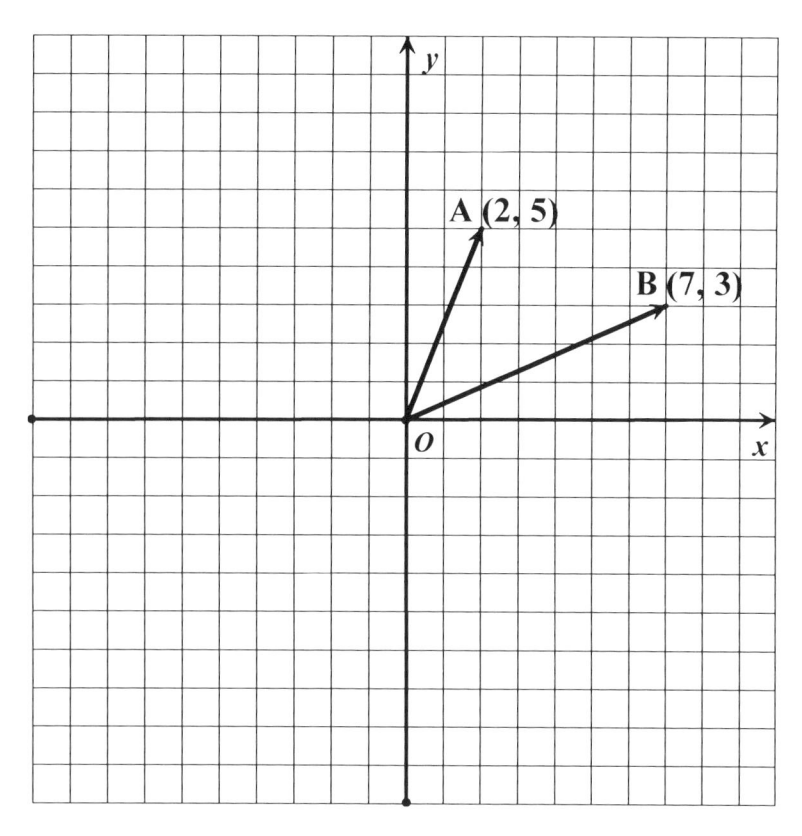

In the figure above, when $\overrightarrow{OA}$ is subtracted from $\overrightarrow{OB}$, what is the length of the resultant vector?

(A) 2.2

(B) 4.0

*(C) 5.4

The resultant of $\overrightarrow{OB} - \overrightarrow{OA} = (7, 3) - (2, 5) = (7 - 2, 3 - 5) = (5, -2)$

The length of the resultant = $|\overrightarrow{OB} - \overrightarrow{OA}| = \sqrt{5^2 + (-2)^2}$
$= \sqrt{25 + 4} = \sqrt{29} \approx 5.4$

Number and Operations; Vectors

(D) 5.8

(E) 7.6

14. The number tau, τ, is defined as two times pi, π. What is the degree measure of an angle whose radian measure is $\dfrac{5\tau}{8}$?

(A)   56.25°

(B)   112.5°

(C)   135°

*(D)   225°

$\tau = 2\pi$

$\pi = \tau/2$

$1° = \left(\dfrac{\pi}{180}\right) radians = \left(\dfrac{\tau/2}{180}\right) radians = \left(\dfrac{\tau}{360}\right) radians$

$\left(\dfrac{5\tau}{8}\right) radians / \left(\dfrac{\tau}{360}\right) radians = 360\left(\dfrac{5}{8}\right)° = 225°$

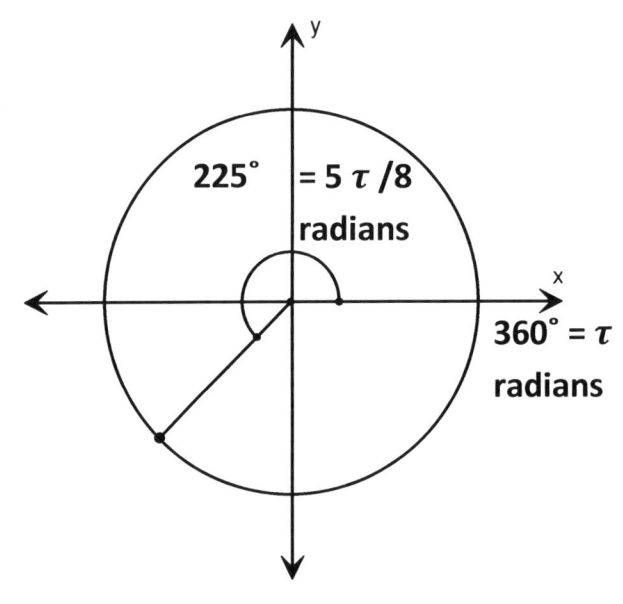

Geometry and Measurement; Trigonometry, Radian Measure

(E)   270°

15. If $\cos \theta = 0.974$, then $\cos(\pi + \theta) =$

*(A)   −0.974

Trigonometric Identities:

$$\sin(\alpha + \beta) = \sin \alpha \cos \beta + \cos \alpha \sin \beta$$

$$\sin(\alpha - \beta) = \sin \alpha \cos \beta - \cos \alpha \sin \beta$$

$$\cos(\alpha + \beta) = \cos \alpha \cos \beta - \sin \alpha \sin \beta$$

$$\cos(\alpha - \beta) = \cos \alpha \cos \beta + \sin \alpha \sin \beta$$

$\cos(\pi + \theta) = \cos \pi \cos \theta - \sin \pi \sin \theta$

$\cos \pi = -1$

$\sin \pi = 0$

$\cos(\pi + \theta) = \cos \pi \cos \theta - \sin \pi \sin \theta$

$= (\cos \theta)(-1) - (\sin \theta)(0) = -\cos \pi$

If $\cos \theta = 0.974$, then $\cos(\pi + \theta) = -0.974$

arccos $0.974 = 0.228532$

$\cos(0.228532 + \pi) = \cos(3.370125) = -0.974$

Geometry and Measurement; Trigonometry, Identities

(B)   −0.228

(C)   0.228

(D)   0.570

(E)   0.974

16. Evaluate *arcsin* (*sin* (π / 2)) in radians.

   (A)   0

   (B)   1

   *(C) π / 2

           sin (π / 2) = 1

           sin θ = 1 when θ = π / 2

           arcsin (*sin* (π / 2)) = π / 2

           Algebra and Functions; Properties of Functions, Inverse Trigonometric

   (D) π

   (E)   2 π

17. The average price of the cars on a lot is $37,180. The median is $35,000 and the standard deviation is $7,000. The dealer reduces the price of each car by $1,000. Which of the following statements are true?

   I.   The new mean is $36,180.

   II. The median remains $35,000.

   III The standard deviation remains $7,000.

   (A)   I only

   (B)   I and II only

   (C)   II and III only

   *(D) I and III only

   The new mean = $37,180 − $1,000 = $36,180.

   The new median = $35,000 − $1,000 = $34,000.

   The standard deviation does not change.

   The following example shows the impact on the mean, median, and standard deviation as a result of reducing the price of each car by $1,000

	Before	After
Car 1	$30,000	$29,000
Car 2	$30,000	$29,000
Car 3	$35,000	$34,000
Car 4	$45,000	$44,000
Car 5	$45,900	$44,900
Mean	$37,180	$36,180
Median	$35,000	$34,000
Std. Dev.	$7,000	$7,000

   Data Analysis, Statistics, and Probability; Mean, Median, and Standard Deviation

   (E)   I, II, and III

18. Which of the following graphs represents the function

$f(x) = 0.4 \cos(3x - 2\pi) + 1$?

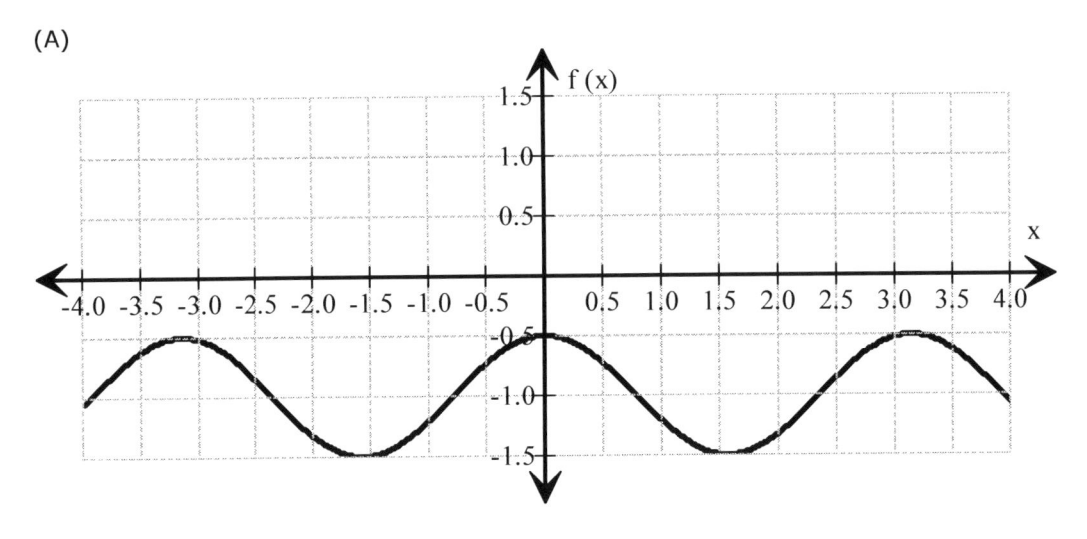

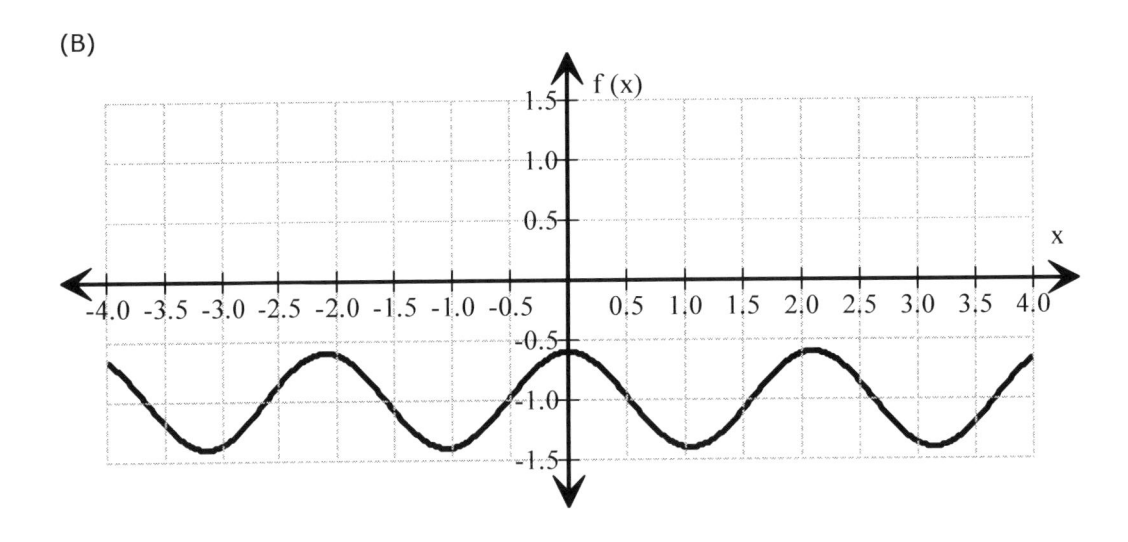

(A)

(B)

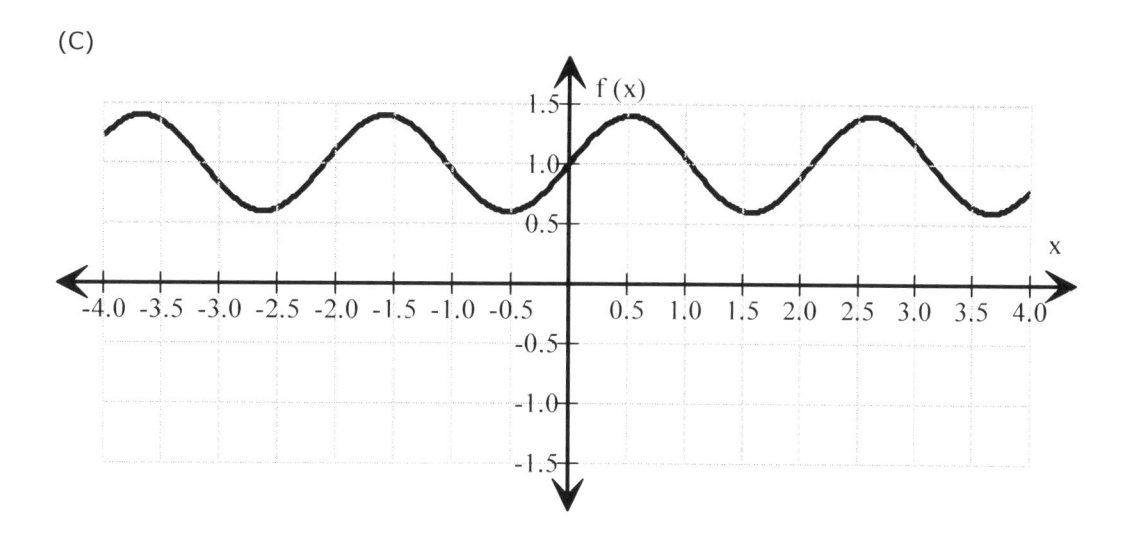

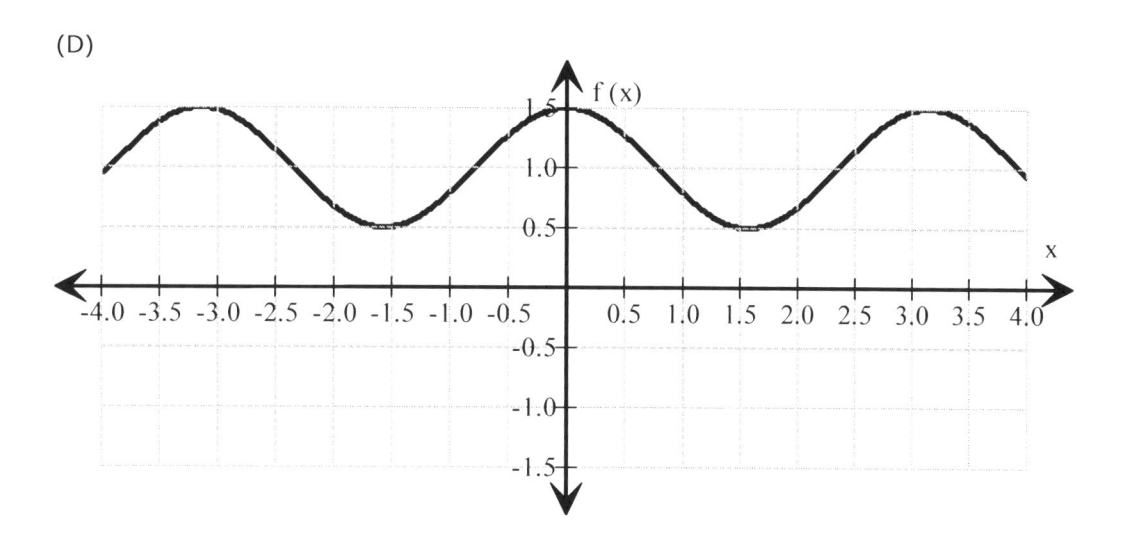

(C)

(D)

*(E)

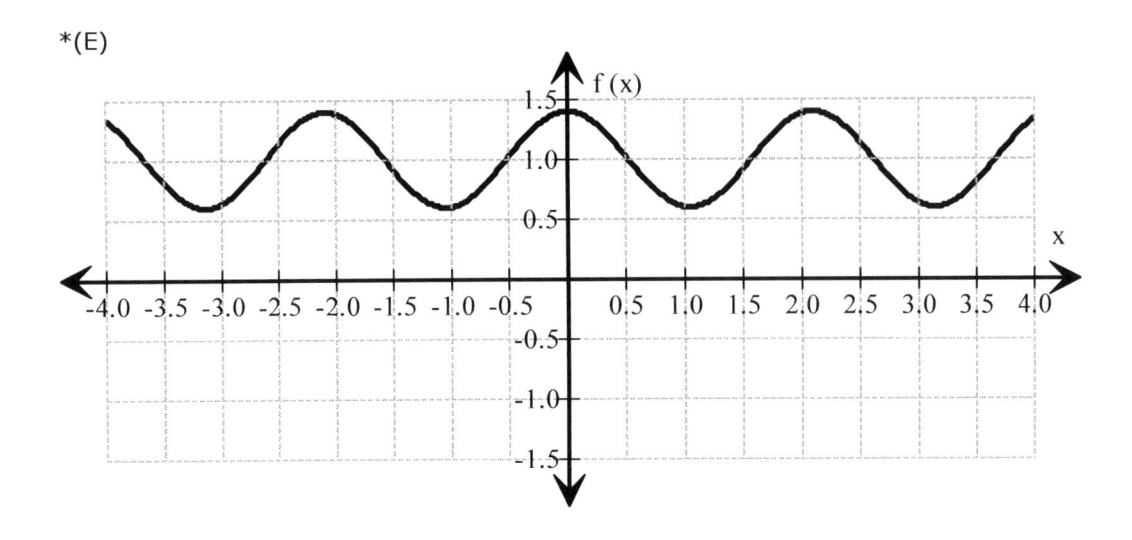

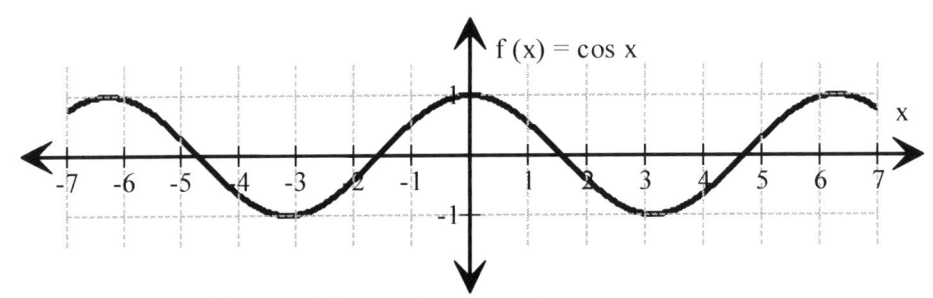

The general form of the cosine function is:

$f(x) = A \cos(Bx + C) + D$

Amplitude A = the distance from the midpoint to the highest or lowest point of the function.

Period T = the distance between any two repeating points on the function.

The period of the cosine function is: $T = 2\pi / |B|$

D = Vertical displacement, or 'y' shift, of the midpoint of the function above the x-axis

Phase shift = the amount of horizontal displacement of the function from its original position.

Phase shift = C / B

$f(x) = 0.4 \cos(3x - 2\pi) + 1$

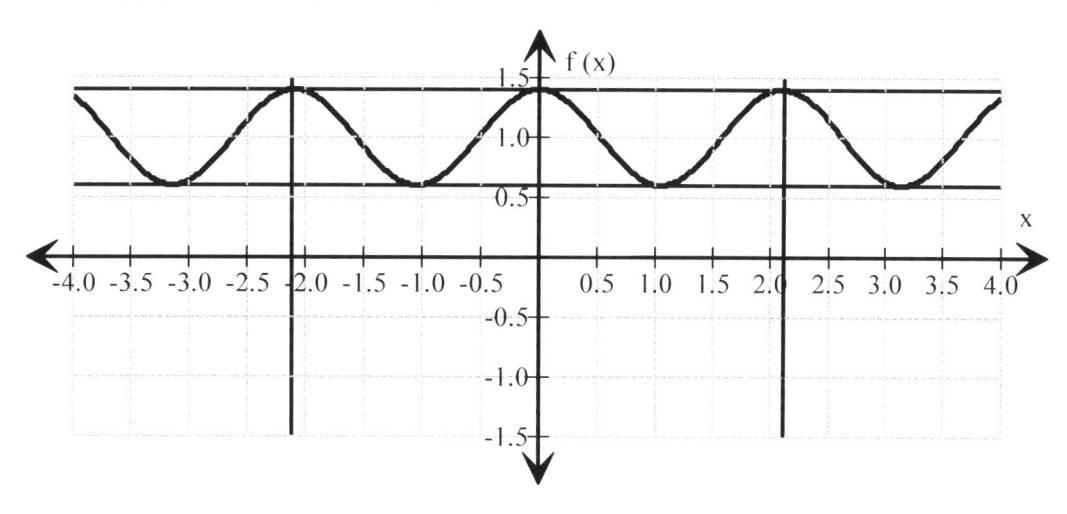

$A = 0.4$

$T = 2\pi/3 = 2\pi/|B|$

$B = 3$

$D = 1$

$C = 2\pi$

$C/B = 2\pi/3$

Algebra and Functions; Properties of Functions, Trigonometric

19. In a group of 30 people, 10 percent are left-handed, and the rest are right-handed. Three people are to be selected at random from the group. What is the probability that all three will be right-handed?

*(A)   0.72

The probability that the first person selected is right-handed

$= (30 - 10\%(30))/30 = 27/30 = 0.9$

The probability the second person selected is right-handed

$= (27 - 1)/(30 - 1) = 26/29$

The probability the third person selected is right-handed

$= (26 - 1)/(29 - 1) = 25/28$

The probability that all three are right-handed

$= (0.9)(26/29)(25/28) = 0.72$

Data Analysis, Statistics, and Probability; Probability

(B)   0.73

(C)   0.81

(D)   0.89

(E)   0.90

20. The formula $N = N_0 / \ln(4t)$ gives the population of a decaying colony of bacteria after t hours for an initial population of N_0, where t is at least equal to 1. How many hours will it take for the population to decrease by 70%?

(A) 1

(B) 2

(C) 4

(D) 6

*(E) 7

$$(1 - 70\%) N_0 = N_0 / \ln(4t)$$

$$1 / 0.3 = \ln(4t)$$

$$e^{\left(\frac{1}{0.3}\right)} = 4t$$

$$t = e^{\left(\frac{1}{0.3}\right)} / 4 = 7$$

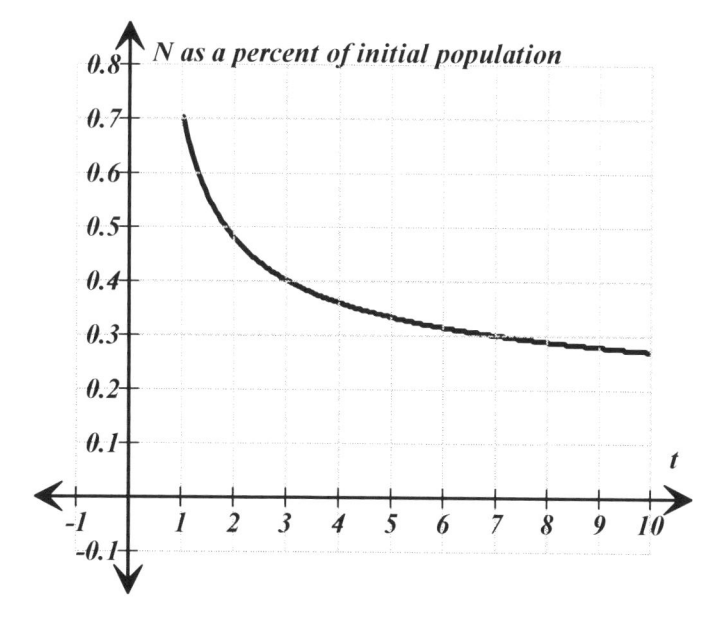

Algebra and Functions; Properties of Functions, Logarithmic

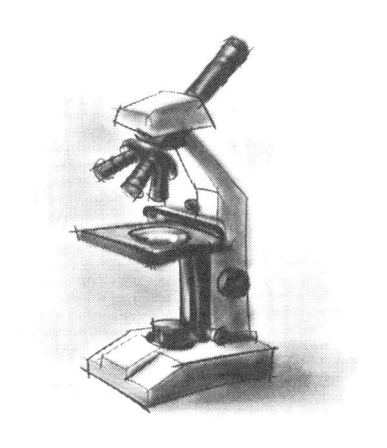

21. For some positive integer t, the first three terms of a geometric sequence are $5t + 1$, $1 - 3t$, and $2t - 2$. What is the numerical value of the fourth term?

(A) −4

*(B) −2

$(2t - 2) / (1 - 3t) = (1 - 3t) / (5t + 1)$

$(2t - 2)(5t + 1) = (1 - 3t)(1 - 3t)$

$10t^2 + 2t - 10t - 2 = 1 - 6t + 9t^2$

$t^2 - 2t - 3 = 0$

$(t - 3)(t + 1) = 0$

$t = 3$ or -1

t is a positive integer

$t = 3$

$5t + 1, 1 - 3t, 2t - 2, ...$

$16, -8, 4, -2, ...$

The fourth term is $(4)(-1/2) = -2$

Number and Operations; Sequences

(C) 0

(D) 2

(E) 4

22. An initial investment of $1,000 increases exponentially to $1,500 after 5 years. At the same rate, in how many years will the value be at least $4,500?

(A) 15

(B) 17

(C) 18

*(D) 19

$$1{,}500 = 1{,}000 \ e^{(5r)}$$

$$e^{(5r)} = 1.5$$

$$\ln e^{(5r)} = \ln 1.5$$

$$5r = 0.405465$$

$$r = \frac{0.405465}{5} = 0.081093$$

$$4{,}500 = 1{,}000 \ e^{0.081093\,t}$$

$$e^{0.081093\,t} = \frac{4{,}500}{1{,}000} = 4.5$$

$$\ln e^{0.081093\,t} = \ln 4.5$$

$$0.081093\,t = \ln 4.5$$

$$= 1.5040774$$

$$t = \frac{1.5040774}{0.081093} = 18.548$$

It will take 19 years for the value to be at least $4,500.

Algebra and Functions; Properties of Functions, Exponential

(E) 20

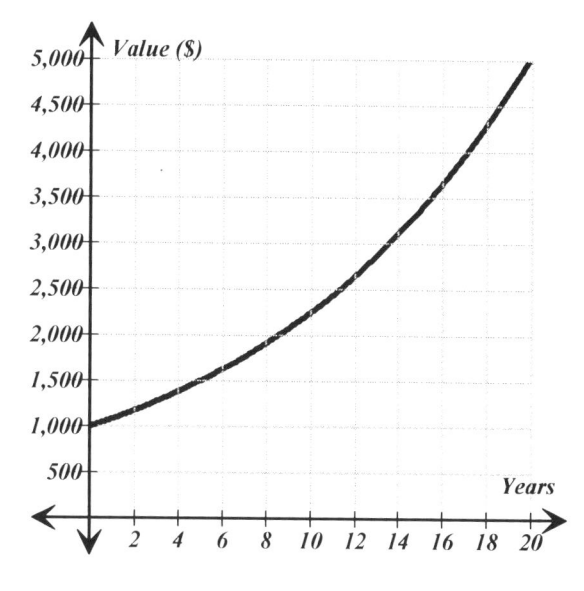

23. If $0 \leq \theta < \pi/2$ and $\sin \theta = 3 \cos \theta$, what is the value of θ ?

(A) 0.24

(B) 0.32

(C) 1.11

*(D) 1.25

$\sin \theta = 3 \cos \theta$

$\sin \theta / \cos \theta = \tan \theta = 3$ when $\theta \neq \pi/2$

$\theta = \tan^{-1} (3) = 1.249046 \approx 1.25$

Geometry and Measurement; Trigonometry, Equations

(E) 1.33

24. What is the range of the function defined by $f(x) = \dfrac{1}{(x-2)(x+1)}$?

(A) All real numbers.

(B) All real numbers except -1 and 2.

*(C) All real numbers greater than 0 or less than or equal to $-4/9$.

$f(x)$ is a rational function with two vertical asymptotes at $x = -1$ and $x = 2$.

x	-3.0	-2.0	-1.1	-0.9	0.0	0.5	1.0	1.9	2.1	3.0
f (x)	0.100	0.250	3.226	-3.448	-0.500	-0.444	-0.500	-3.448	3.226	0.250

The range of $f(x)$ is all real numbers greater than 0 or less than or equal to $-4/9$

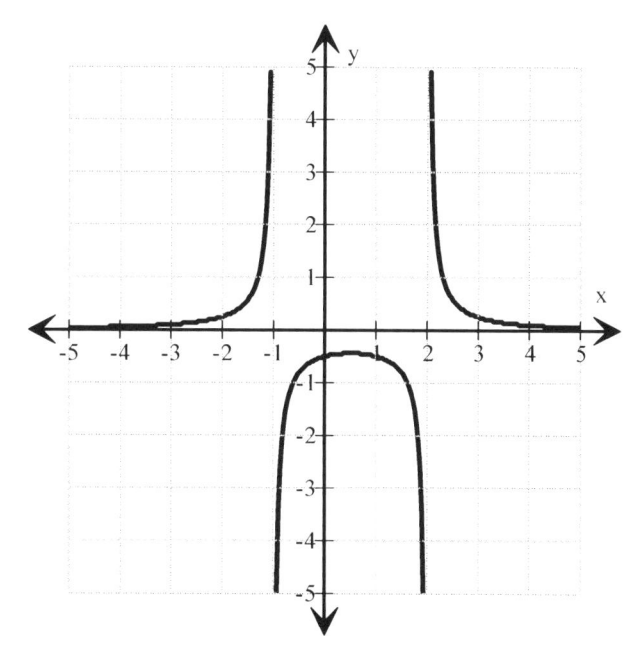

Algebra and Functions; Properties of Functions, Rational

(D) All real numbers greater than 0 or less than or equal to -0.5.

(E) All real numbers greater than 0 or less than -2.25.

25. What is the sum of the infinite geometric series $6 + 2 + \dfrac{2}{3} + \dfrac{2}{9} + \ldots$?

*(A)   9

$$\sum a_i = \frac{a_1}{1-r} = \frac{6}{1 - 1/3} = \frac{6}{2/3} = 9$$

n	1	2	3	4	5	6	7	8	9
a_n	6.000	2.000	0.667	0.222	0.074	0.025	0.008	0.003	0.001
$\sum a_n$	6.000	8.000	8.667	8.889	8.963	8.988	8.996	8.999	9.000

Number and Operations; Series

(B)   12

(C)   13.5

(D)   15

(E)   18

26. Which of the following graphs could be the parametric representation of the equations $x = a \cos(t)$ and $y = a \sin(t)$ where $0 \le t \le 2$?

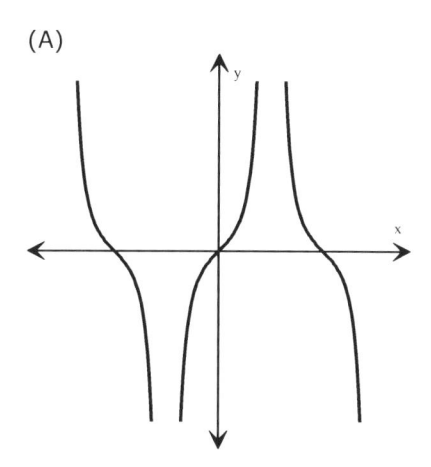

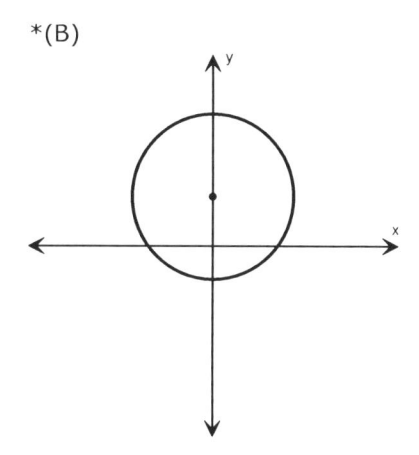

(A)

*(B)

$\dfrac{x}{a} = \cos(t)$

$\dfrac{y}{a} = \sin(t)$

$(\cos(t))^2 + (\sin(t))^2 = 1$

$(x/a)^2 + (y/a)^2 = 1$

The equations are the parametric representation of a circle.

Algebra and Functions; Properties of Functions, Parametric

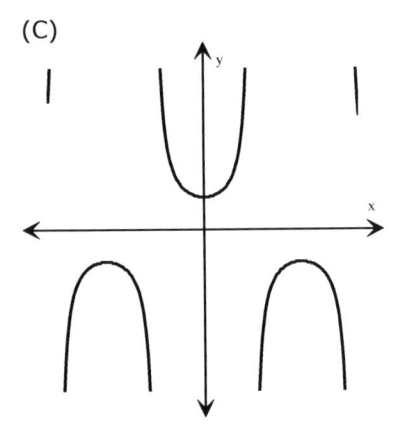

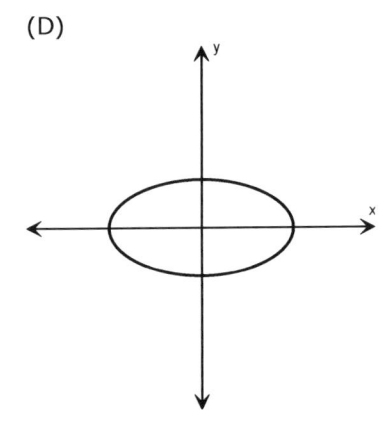

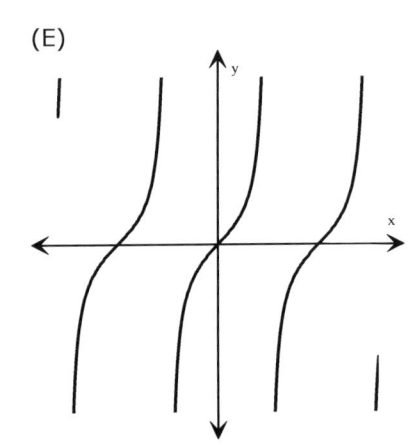

(C)

(D)

(E)

27.

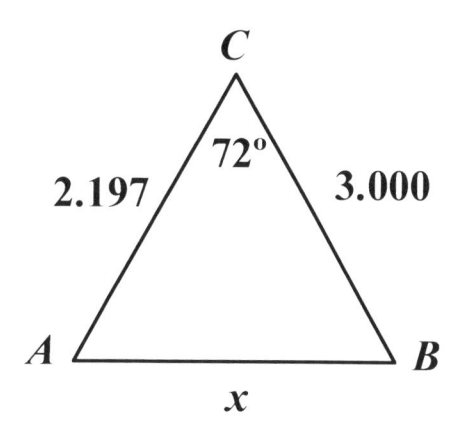

Note: Figure not drawn to scale.

In the figure above, AC = 2.197, BC = 3.000, and ∠ C = 72°. What is the value of x ?

(A)   2.53

(B)   2.83

(C)   3.08

*(D) 3.12

Law of Cosines: $c^2 = a^2 + b^2 - 2\,a\,b\,(\cos C)$

$x^2 = 3.0^2 + 2.197^2 - 2\,(3.0)(2.197)\,(\cos 72°)$

$x^2 \approx 9 + 4.826809 - 13.182\,(0.309016994) \approx 13.826809 - 4.073462$
$= 9.753347$

$x \approx \sqrt{9.753347} \approx 3.123$

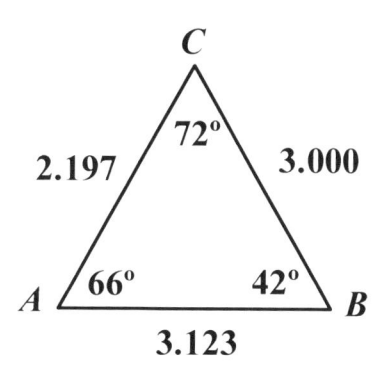

Triangle *ABC* can be split into 2 right triangles *ACD* and *BCD*.

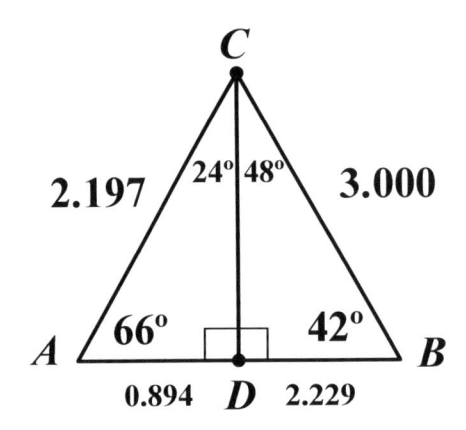

∠ *ACD* = 180° − 90° − 66° = 24°

sin 24° = *AD* / 2.197

0.406736643 = *AD* / 2.197

AD = (0.406736643) (2.197) ≈ 0.894

∠ *BCD* = 180° − 90° − 42° = 48°

sin 48° = *DB* / 3.000

0.743144825 = *DB* / 3.000

AD = (0.743144825) (3.000) ≈ 2.229

AB = *AD* + *DB* = 0.894 + 2.229 = 3.123

Law of Sines: sin *A* / *BC* = sin *B* / *AC* = sin *C* / *AB*

sin *A* / 3.0 = sin *B* / 2.2 = sin 72° / *x*

Angle	A	C	C
Opposite Side	BC	AB	AB
Measure (degrees)	66	72	42
sin	0.914	0.951	0.669
Opposite side length	3.000	3.123	2.197
sin / opposite side	0.305	0.305	0.305

Geometry and Measurement; Trigonometry, Law of Cosines

(E)   3.28

28. The Fibonacci sequence is recursively defined by $a_n = a_{n-1} + a_{n-2}$, for $n > 2$. If $a_1 = 1$ and $a_2 = 1$, what is the value of a_{17}?

(A) 610

(B) 987

*(C) 1,597

n	1	2	3	4	5	6	7	8	9	10
$a_{n-1} + a_{n-2}$	1	1	2	3	5	8	13	21	34	55

n	11	12	13	14	15	16	17	18	19	20
$a_{n-1} + a_{n-2}$	89	144	233	377	610	987	1,597	2,584	4,181	6,765

Algebra and Functions; Properties of Functions, Recursive

(D) 2,584

(E) 4,181

29.

800, 300, 770, 710, 700, 650, 650, 600, 560, 760

A group of students had the above scores on a college entrance exam. If the percentiles are in the range 0 to 100, inclusive, what is the interquartile range for the scores?

(A)   60

*(B) 135

Including the percentiles at 0 and 100, the 25th percentile out of n terms is a weighted average of terms to get the term in $[(n + 3) / 4]^{th}$ place.

Order	1	2	3	4	5	6	7	8	9	10
Return	800	770	760	710	700	650	650	600	560	300
Percentile	100	89	78	67	56	44	33	22	11	0

For $n = 10$, $[(n + 3) / 4] = (10 + 3) / 4 = 13 / 4 = 3.25$

The term in place $3.25 = (0.75)(760) + (0.25)(710) = 747.5$

Including the percentiles at 0 and 100, the 75th percentile out of n terms is a weighted average of terms to get the term in $[(3n + 1) / 4]^{th}$ place.

For $n = 10$, $[(3n + 1) / 4] = (30 + 1) / 4 = 31 / 4 = 7.75$

The term in place $7.75 = (0.25)(650) + (0.75)(600) = 612.5$

The interquartile range is the absolute value of the difference between the 25th percentile and the 75th percentile.

$747.5 - 612.5 = 135.0$

Excluding the percentiles at 0 and 100, the interquartile range would be $762.5 - 590 = 172.5$.

Data Analysis, Statistics, and Probability; Interquartile Range

(C)   160

(D)   173

(E)   185

30. What annual compounded rate of return would result in an initial investment of $1,000 accumulating to $4,000 in 18 years?

(A) 7.2%

(B) 7.3%

(C) 7.7%

*(D) 8.0%

$$(1+i)^{18} = \frac{\$4,000}{\$1,000} = 4$$

$$18 \ln(1+i) = \ln 4$$

$$\ln(1+i) = \frac{\ln 4}{18} \approx 0.077063$$

$$e^{0.077063} \approx 1.080 = 1+i$$

$$i = 0.080 = 8.0\%$$

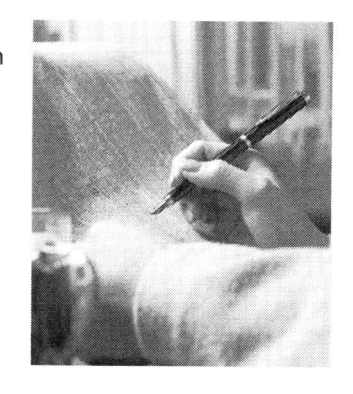

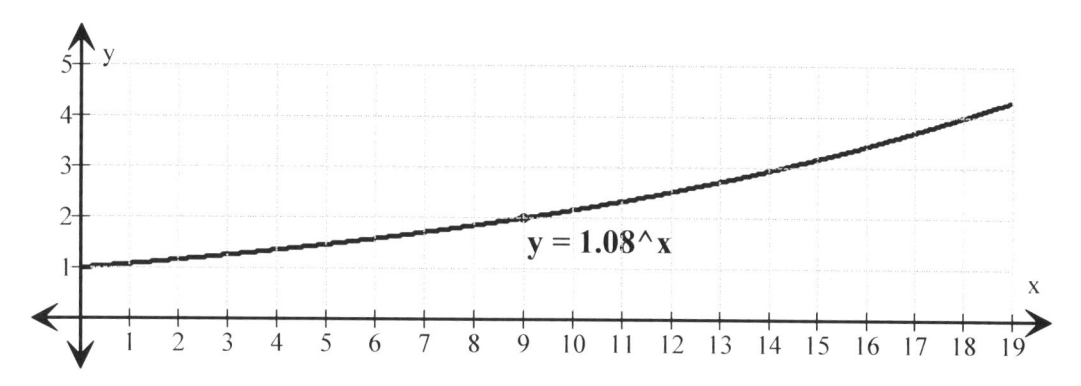

$y = 1.08^\wedge x$

Algebra and Functions; Properties of Functions, Exponential

(E) 8.6%

31. Find the exponential function that best fits the data below.

x	0.1	1.2	1.9	3.1	4.2	4.9
y	0.60	0.70	0.90	1.15	1.42	1.82

(A) $y = (0.57)(1.24^x)$

(B) $y = (0.57)(1.25^x)$

*(C) $y = (0.57)(1.26^x)$

Using the TI-83 Plus calculator:

STAT

ENTER

L1(1) = 0.1, L1(2) = 1.2, L1(3) = 1.9,
L1(4) = 3.1, L1(5) = 4.2, L1(6) = 4.9

L2(1) = 0.6, L2(2) = 0.7, L2(3) = 0.9,
L2(4) = 1.15, L2(5) = 1.42, L2(6) = 1.82

STAT

CALC

0

ExpReg

ENTER

The screen should show the following:

ExpReg
  y = a * b ^ x
  a = 0.5650119636
  b = 1.2585887299

$y = (0.57)(1.26^x)$

Data Analysis, Statistics, and Probability; Least Squares Regression, Exponential

(D) $y = (0.58)(1.25^x)$

(E) $y = (0.58)(1.26^x)$

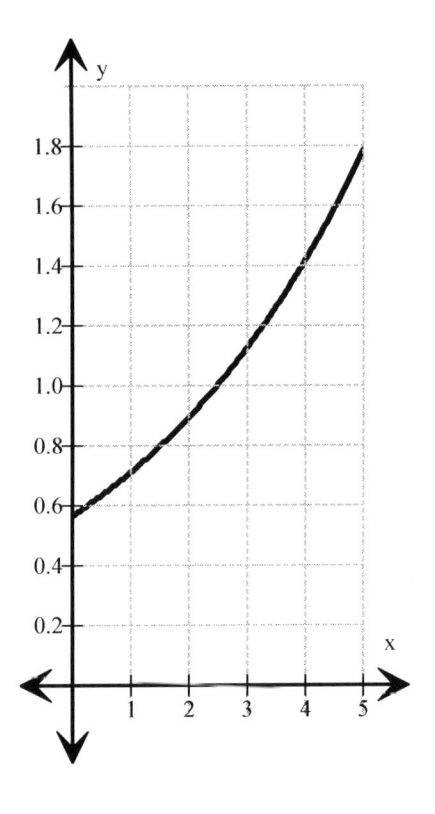

32. If $f(x) = \sqrt[4]{\dfrac{x^2}{3} + 2}$, what is $f^{-1}(1.8)$?

(A) 1.05

(B) 1.79

(C) 2.91

(D) 4.70

*(E) 5.05

$$f^{-1}(1.8) = [((1.8)^4 - 2)(3)]^{(1/2)} = 5.05$$

$$f(5.05) = [((5.05)^2/3 + 2)]^{(1/4)} = 1.8$$

Algebra and Functions; Properties of Functions, Exponential

33. Using the quadratic regression model, which parabola best fits the data for the number of new homes sold in a community?

Year	0	1	2	3	4	5	6
Number Sold	98	76	58	54	60	74	99

(A) $y = -4.8 x^2 + 29.5 x - 98.6$

(B) $y = -4.7 x^2 + 29.4 x - 98.5$

(C) $y = 4.7 x^2 - 29.4 x + 98.5$

(D) $y = 4.8 x^2 - 29.5 x + 98.6$

*(E) $y = 4.9 x^2 - 29.6 x + 98.7$

Using the TI-83 Plus calculator:

STAT
ENTER

L1(1) = 0, L1(2) = 1, L1(3) = 2, L1(4) = 1, L1(5) = 4, L1(6) = 5, L1(7) = 6

L2(1) = 98, L2(2) = 76, L2(3) = 58, L2(4) = 54, L2(5) = 60, L2(6) = 74, L2(7) = 99

STAT
CALC
5
QuadReg
ENTER

The screen should show the following:

ExpReg
$y = a x^2 + b x + c$
a = 4.94047619
b = -29.60714286
c = 98.73809524

$y = 5 x^2 - 30 x + 99$

Data Analysis, Statistics, and Probability; Least Squares Regression, Quadratic

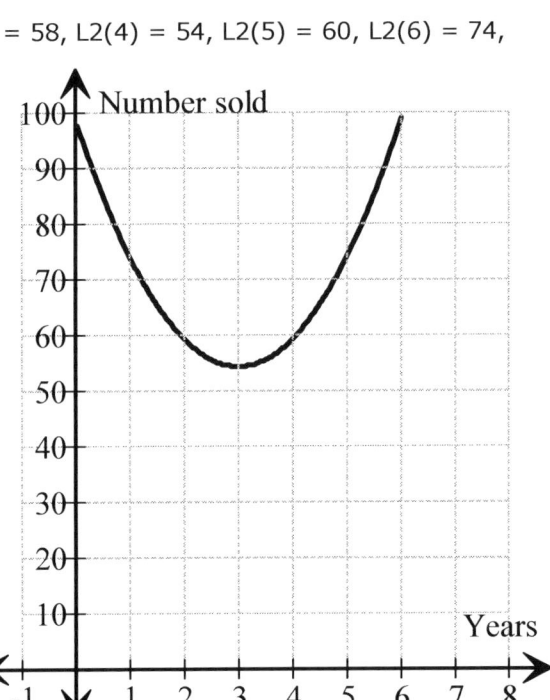

34. What value does $\dfrac{2\ln(x)}{1-x}$ approach as x approaches 1?

*(A)   −2

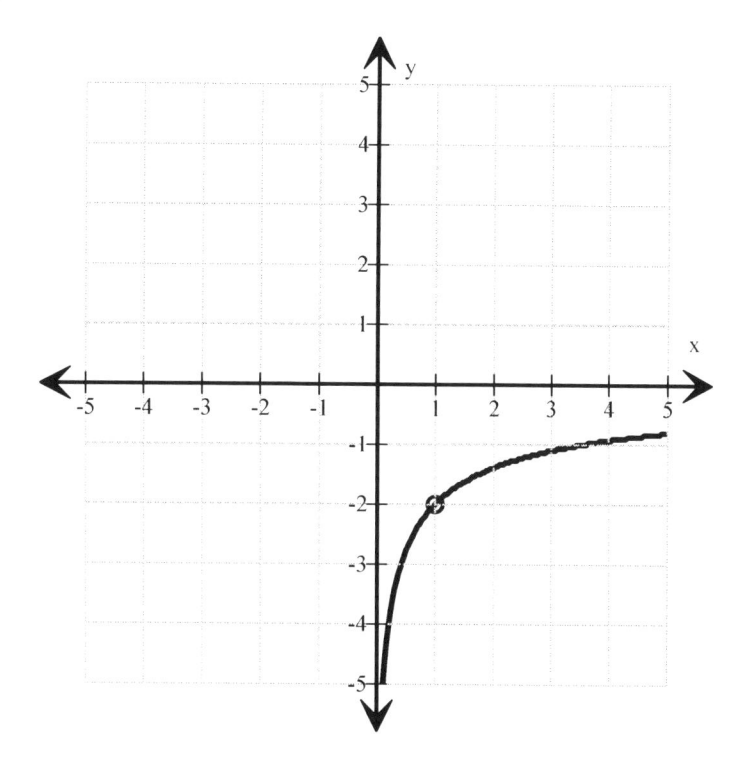

The value of ln (x) / (1 − x) approaches −2 as x approaches 1 from both sides.

2 (ln 1.01) / (1 − 1.01) = −1.99

2 (ln 0.99) / (1 − 0.99) = −2.01

$\lim_{x \to 1} \left(\dfrac{2\ln(x)}{1-x} \right) = -2$

Algebra and Functions; Properties of Functions, Logarithmic

(B)   −1

(C)   0

(D)   1

(E)   2

35. Which of the following statements are accurate regarding symmetry of functions?

   I.   There is symmetry about the x-axis if $f(x) = f(-x)$

   II. There is symmetry about the y-axis if $f(x) = -f(x)$

   III. There is symmetry about the origin if $f(x) = -f(-x)$

(A)   I only

(B)   II only

*(C) III only

Symmetry about the x-axis if $f(x) = -f(x)$

Symmetry about the y-axis if $f(x) = f(-x)$

Symmetry about the origin if $f(x) = -f(-x)$

Geometry and Measurement; Symmetry

(D)   I and II only

(E)   I and III only

36. The function f is defined by $f(x) = \dfrac{x^3}{3} - \dfrac{3x^2}{4} - \dfrac{x}{2} + 2$ for $-1 \leq x \leq 2$. What is the difference between the maximum and minimum values of (x)?

(A)   1.33

(B)   1.41

(C)   1.42

*(D)   1.46

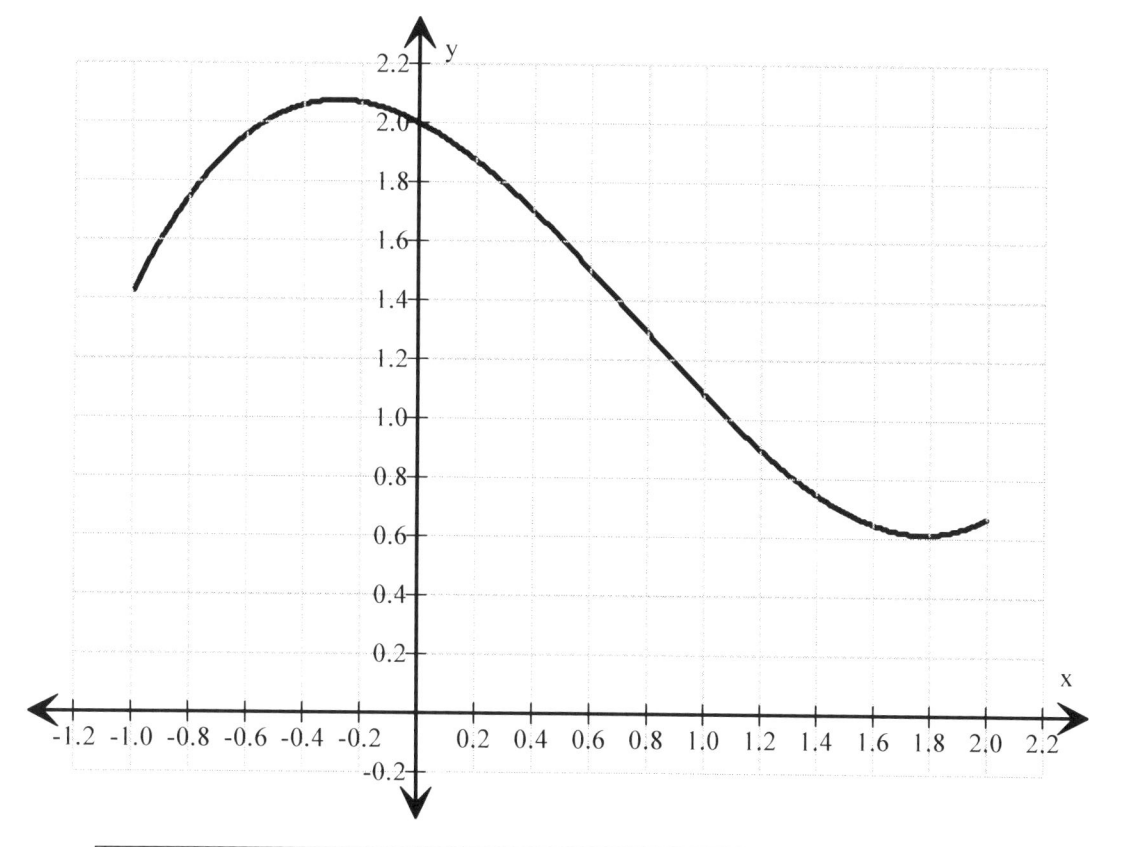

x	-1.000	-0.280	0.000	1.000	1.780	2.000
f(x)	1.417	2.074	2.000	1.083	0.614	0.667

2.074 - 0.614 = 1.46

Algebra and Functions; Properties of Functions, Polynomial

(E)   1.50

37.

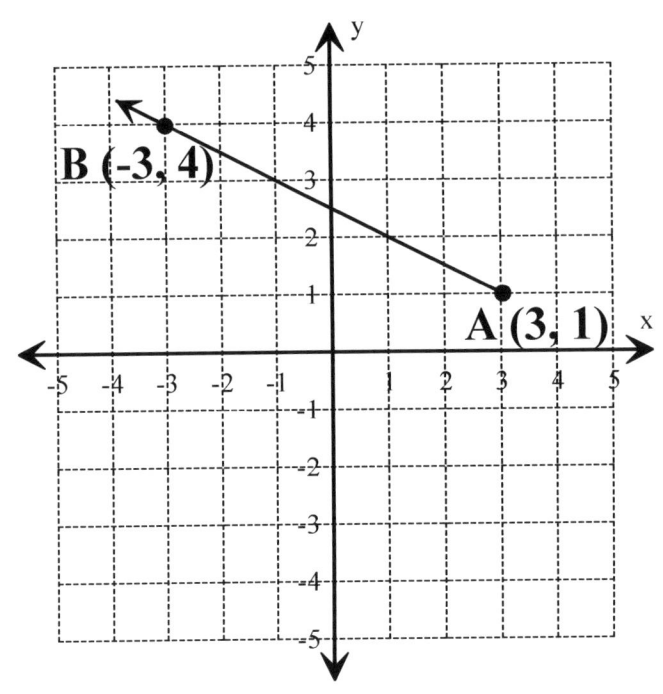

Which of the following vector equations is represented by the above graph, given that an object moves from point A to point B in 2 seconds?

*(A) $y = (3, 1) + t(-3, 1.5)$

The object starts at point (3, 1) when $t = 0$.

The object is at point (−3, 4) when $t = 2$.

$x = 3 + t\left(\frac{-3-3}{2}\right) = 3 - 3t$

$y = 1 + t\left(\frac{4-1}{2}\right) = 1 + 1.5t$

$y = (3, 1) + t(-3, 1.5)$

Algebra and Functions; Properties of Functions, Parametric

(B) $y = (3, 1) + t(-2, 1)$

(C) $y = (3, 1) - t(1, -0.5)$

(D) $y = (5, 0) + t(-4, 2)$

(E) $y = (5, 0) - t(4, 2)$

38. What is the period of the mantissa function, $(x) = x - [x]$, where $[x]$ is the greatest integer less than or equal to x?

(A) -1

(B) 0

*(C) 1

The mantissa is the decimal part of a number.

For example, the mantissa of 2.345 is 0.345.

The mantissa of an integer is 0.

The decimal part of a number is at least 0 and less than 1.

$0 \leq x - [x] < 1$

The mantissa function has a period of 1.

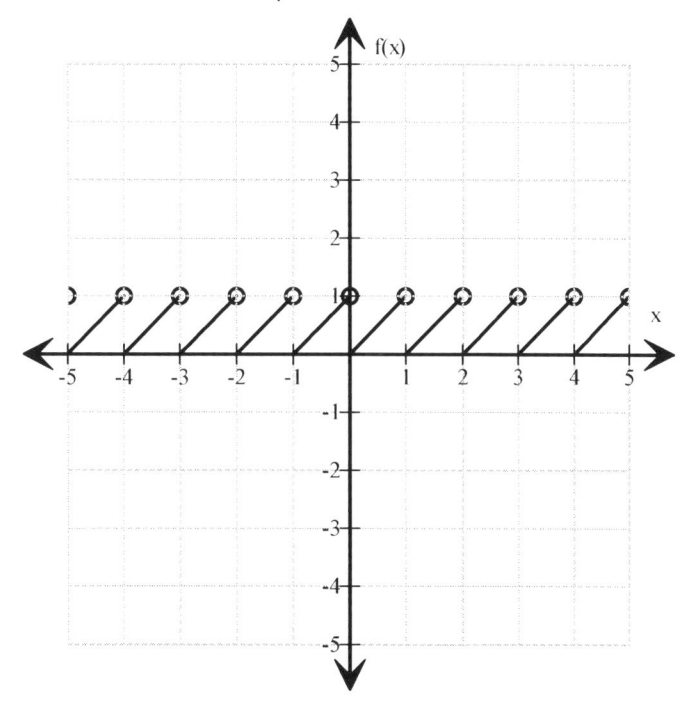

Algebra and Functions; Properties of Functions, Piecewise

(D) $\sqrt{2}$

(E) 2

39.

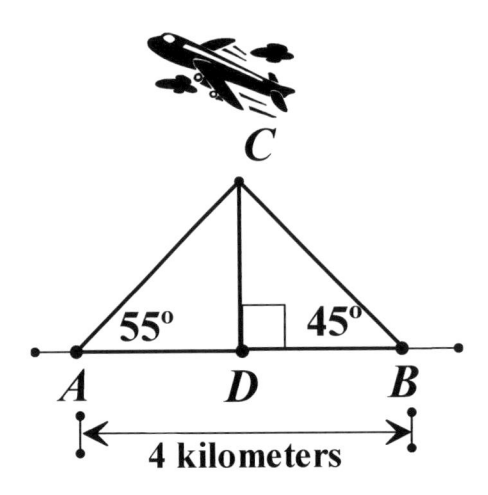

The airplane in the figure above is directly over point D at point C on a level runway. The angles of elevation for points A and B are 55° and 45°, respectively. If points A and B are 4 kilometers apart, what is the distance, in kilometers, from point B to the airplane?

(A)   1.65

(B)   2.35

(C)   2.83

(D)   2.87

*(E)   3.33

Law of Sines: sin A / BC = sin B / AC = sin C / AB

$\angle C = 180° - 55° - 45° = 80°$

$\sin 80° / 4 = \sin 55° / BC$

$BC = 4 \sin 55° / \sin 80° = (4)(0.81915) / (0.98481) = 3.327$

Angle	A	B	C
Opposite Side	BC	AC	AB
Measure (degrees)	55	45	80
sin	0.819151	0.70711	0.98481
Opposite side length	3.327	2.872	4.000
sin / opposite side	0.2462	0.2462	0.2462

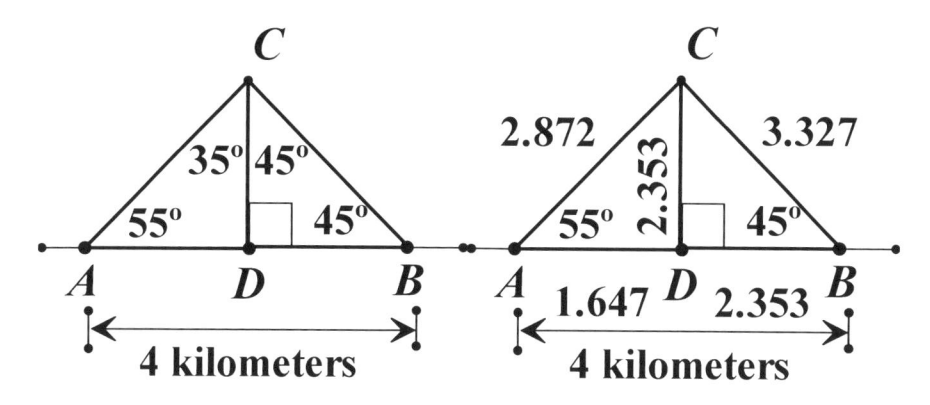

$DB^2 + CD^2 = CB^2 = (3.327)^2 = 11.0689$

$AD^2 + CD^2 = AC^2 = (2.872)^2 = 8.2484$

$DB^2 - AD^2 = 11.0689 - 8.2484 = 2.8205$

$AD + DB = 4$

$AD = 4 - DB$

$AD^2 = 16 - 8\,DB + DB^2$

$DB^2 - (16 - 8\,DB + DB^2) = 2.8205$

$8\,DB - 16 = 2.8205$

$8\,DB = 18.8205$

$DB = 18.8205 / 8 = 2.353$

$CD = DB = 2.353$

$AD = 4 - DB = 4 - 2.353 = 1.647$

$CB = \sqrt{2}\,DB = \sqrt{2}\,CD$

$3.327 = \sqrt{2}\,DB = \sqrt{2}\,CD$

$DB = CD = 3.327 / \sqrt{2} = 2.353$

$AD = 4 - DB = 4 - 2.353 = 1.647$

Geometry and Measurement; Trigonometry, Law of Sines

40. Which of the following graphs is the best parametric representation of the equations $x = -5\cos(t)$ and $= \sin(t)$?

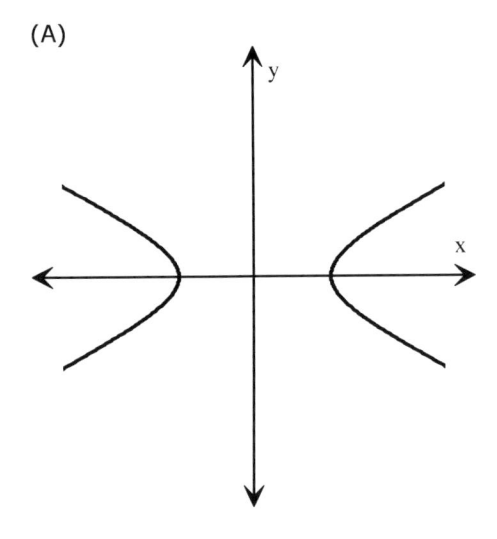

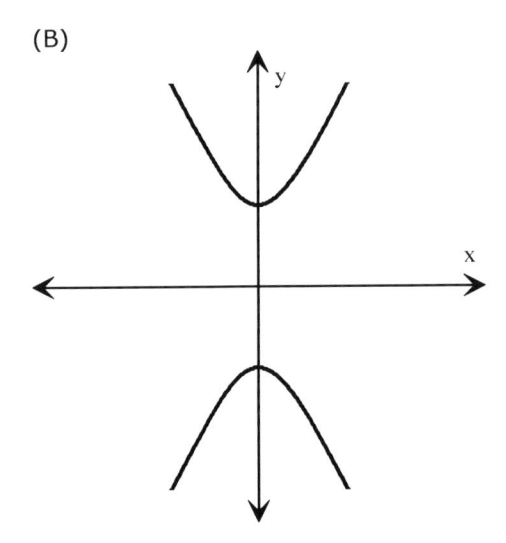

(A)

(B)

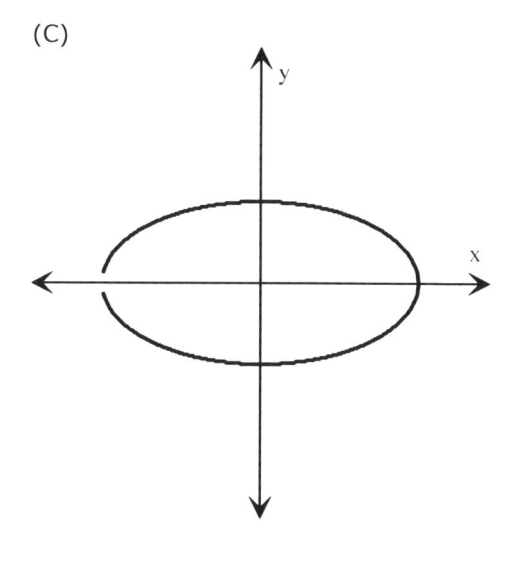

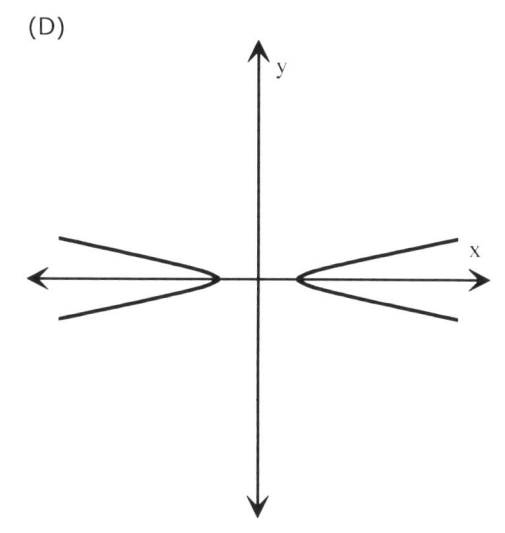

(C)

(D)

*(E)

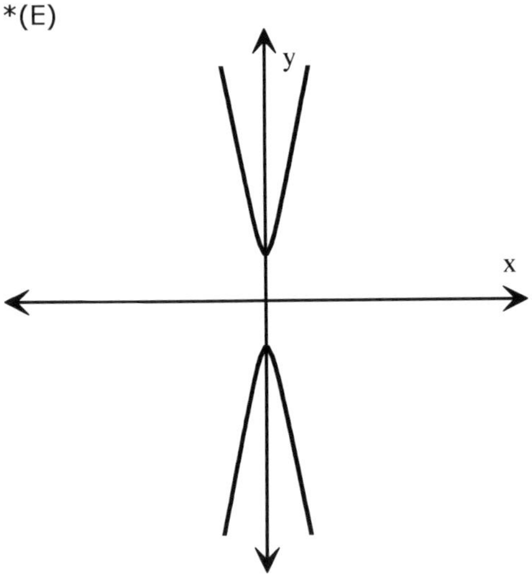

$$\frac{-x}{5} = \cos(t)$$

$$y = \sin(t)$$

$$(cos\ (t))^2 + (\sin(t))^2 = 1$$

$$y^2 - (x/5)^2 = 1$$

The equations are the parametric representation of an up and down hyperbola.

Algebra and Functions; Properties of Functions, Parametric

41. Which of the following expressions is equivalent to $\sin(\pi + 2\theta)$?

(A) $\sin^2\theta$

(B) $\sin\theta \cos\theta$

(C) $-\sin\theta \cos\theta$

(D) $2\sin\theta \cos\theta$

*(E) $-2\sin\theta \cos\theta$

$\sin(\alpha + \beta) = \sin\alpha \cos\beta + \cos\alpha \sin\beta$

Double Angle Formulas:

$\sin 2\alpha = 2\sin\alpha \cos\alpha$

$\cos 2\alpha = \cos^2\alpha - \sin^2\alpha$

$\cos 2\alpha = 1 - 2\sin^2\alpha$

$\cos 2\alpha = 2\cos^2\alpha - 1$

$\cos\pi = -1$

$\sin\pi = 0$

$\sin(\pi + 2\theta) = \sin\pi \cos 2\theta + \cos\pi \sin 2\theta$

$= (0)\cos 2\theta + (-1)\sin 2\theta = -\sin 2\theta = -2\sin\theta \cos\theta$

Let $\theta = 0.25$

$\sin(\pi + (2)(0.25)) = -0.4794$

$-2\sin(0.25)\cos(0.25) = -0.4794$

Geometry and Measurement; Trigonometry, Double Angle Formulas

42.

$$f(x) = 0.5 \sin(nx + \pi) + 1.25$$

For what value of n, where $n > 0$, is the period of the above function equal to 4π ?

(A)   0.25

*(B) 0.50

The general form of the sine function is:

$y = A \sin(Bx + C) + D$

Period T = the distance between any two repeating points on the function.

The period of the sine function is: $T = 2\pi / |B|$

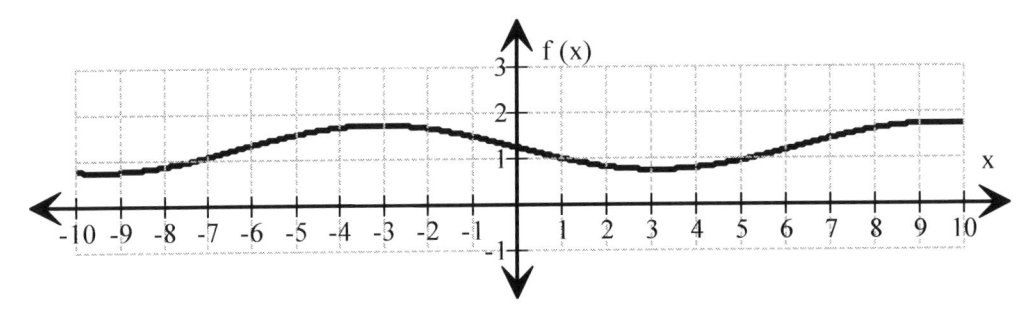

$T = 4\pi = 2\pi / |B|$

$B = 2\pi / 4\pi = 0.5$

Algebra and Functions; Properties of Functions, Periodic

(C)   0.75

(D)   1.00

(E)   1.25

43. Suppose the graph of $f(x) = 1 - x^2$ is translated 2 units right and 3 units down. If the resulting graph is $g(x)$, what is the value of $g(-2)$?

*(A)   −18

$$f(x) = 1 - x^2$$

$$g(x) = -2 - (x-2)^2$$

$$g(-2) = -2 - (-2-2)^2 = -2 - (-4)^2 = -2 - 16 = -18$$

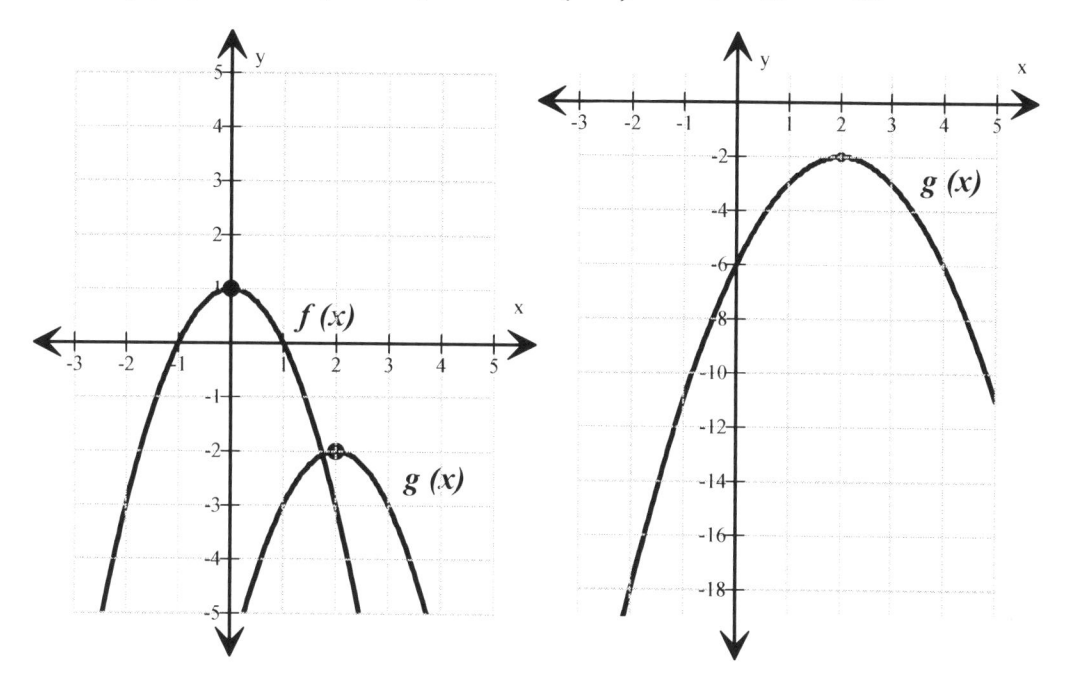

Geometry and Measurement, Coordinate, Parabolas

(B)   −11

(C)   −5

(D)   −2

(E)   3

44.

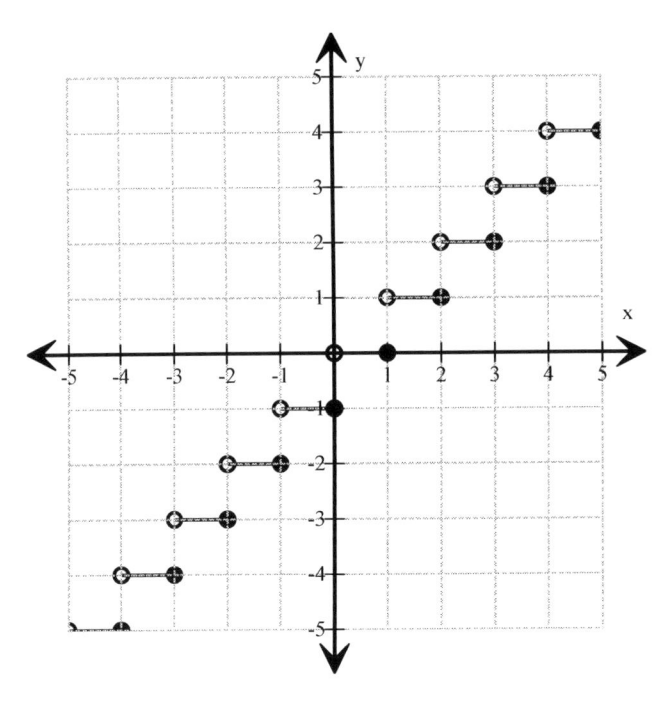

Which of the following expressions best describes the above graph?

(A) The floor function or greatest integer function $y = [x]$ where y is the greatest integer not less than x

(B) The ceiling function where y is the smallest integer not less than x

*(C) The function where y is the smallest integer not less than $x - 1$

x	$x - 1$	y	$y \geq x - 1$
-0.5	-1.5	-1	$-1 \geq -1.5$
0.0	-1.0	-1	$-1 \geq -1$
0.5	-0.5	0	$0 \geq -0.5$
1.0	0.0	0	$0 \geq 0$
1.5	0.5	1	$1 \geq 0.5$

Algebra and Functions; Properties of Functions, Piecewise

(D) The function where y is the greatest integer not less than $x - 1$

(E) The mantissa function $y = x - [x]$

45. Which of the following graphs could be the parametric representation of the equations $x = 2\cos(t)$ and $y = 4\sin(t)$?

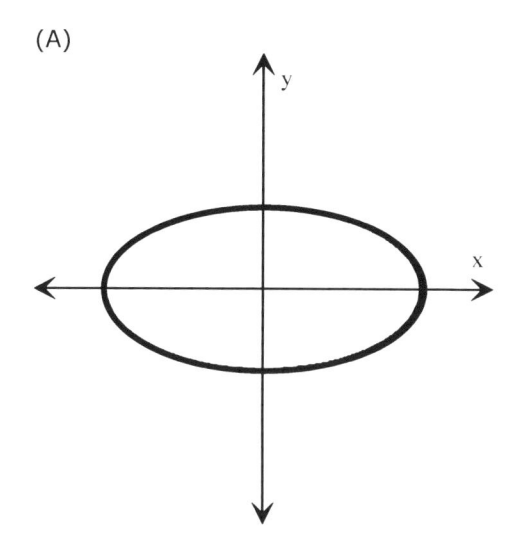

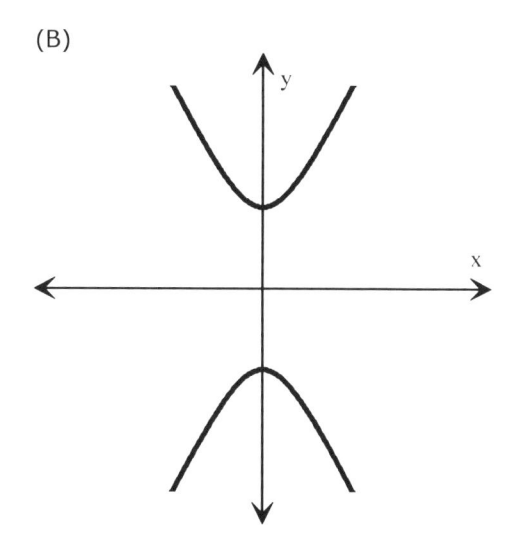

(A)

(B)

*(C)

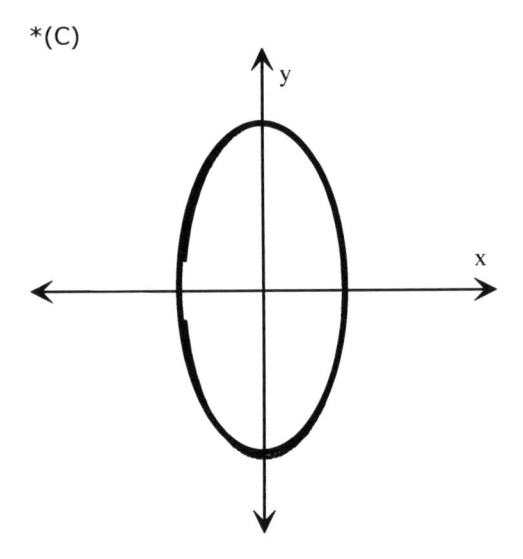

[Ellipse centered at origin with vertical major axis, shown on x-y coordinate plane]

$$\frac{x}{2} = \cos(t)$$

$$\frac{y}{4} = \sin(t)$$

$$(\cos(t))^2 + (\sin(t))^2 = 1$$

$$(x/2)^2 + (y/4)^2 = 1$$

The equations are the parametric representation of an ellipse with a vertical major axis.

Algebra and Functions; Properties of Functions, Parametric

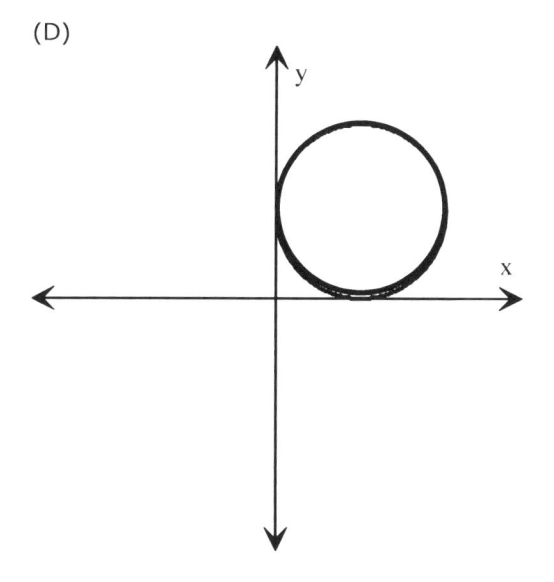

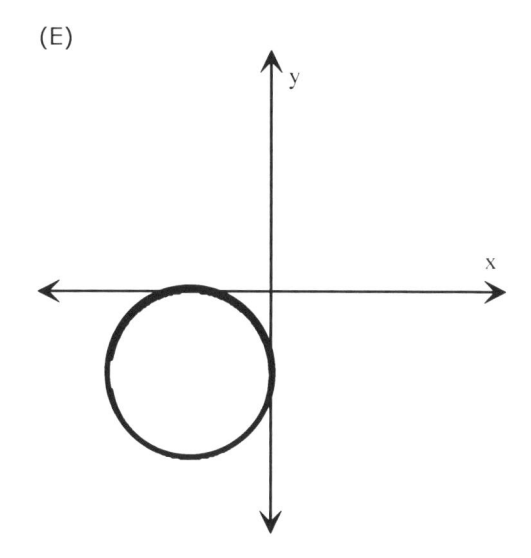

(D)

(E)

46. Express the point with coordinates (2.76, 2.31) in terms of polar coordinates.

(A) (1.5, 40°)

(B) (1.5, 50°)

(C) (3.6, 30°)

*(D) (3.6, 40°)

$x = r \cos \theta$

$y = r \sin \theta$

$x^2 + y^2 = r^2$

$2.76 = r \cos \theta$

$2.31 = r \sin \theta$

$2.76^2 + 2.31^2 = r^2$

$r^2 = 7.6176 + 5.3361 = 12.9537$

$r = 3.6$

$3.6 \cos \theta = 2.76$

$\cos \theta = \dfrac{2.76}{3.6} = 0.7667$

$\theta = 40°$

$(r, \theta) = (3.6, 40°)$

$3.6 \sin 40° = 2.31$

$\sin \theta = \dfrac{2.31}{3.6} = 0.6417$

$\theta = 40°$

Geometry and Measurement; Coordinate, Polar Coordinates

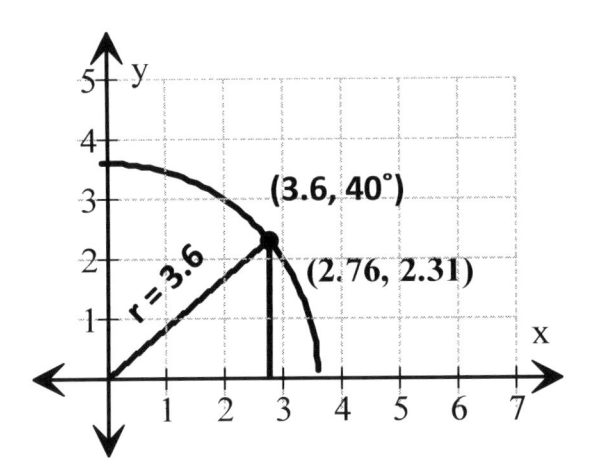

(E) (3.6, 50°)

47. Which of the following statements are true regarding inverse trigonometric functions?

   I.   The domain of $y = \sin^{-1} x$ is $-1 \leq x \leq 1$, and the range is $-\frac{\pi}{2} \leq y \leq \frac{\pi}{2}$.

   II. The domain of $y = \cos^{-1} x$ is $-1 \leq x \leq 1$, and the range is $0 \leq y \leq \pi$.

   III The domain of $y = \tan^{-1} x$ is all real numbers, and the range is $0 \leq y \leq \pi$.

   (A)   I only

   *(B) I and II only

The domain of $y = \sin^{-1} x$ is $-1 \leq x \leq 1$, and the range is $-\frac{\pi}{2} \leq y \leq \frac{\pi}{2}$.

The domain of $y = \cos^{-1} x$ is $-1 \leq x \leq 1$, and the range is $0 \leq y \leq \pi$.

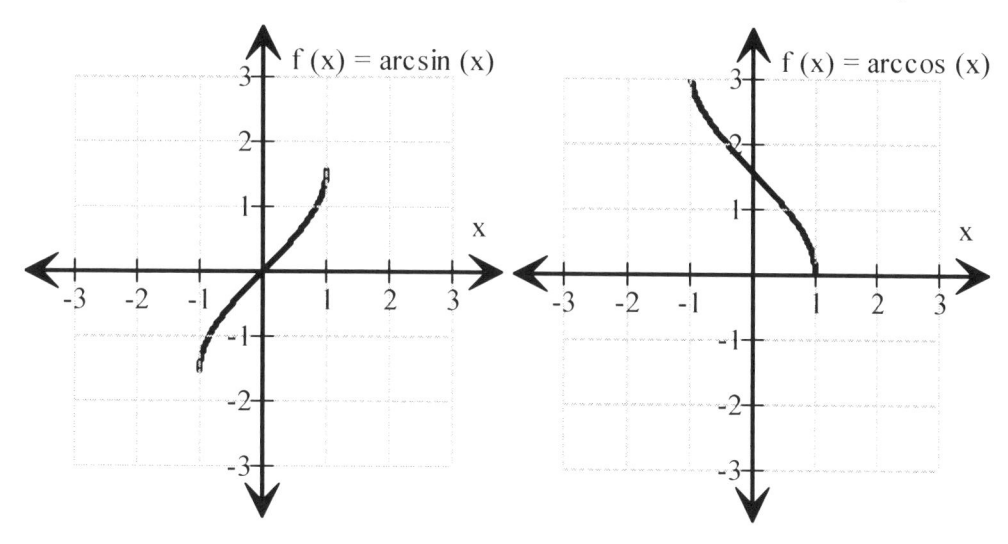

The domain of $y = \tan^{-1} x$ is all real numbers, and the range is $-\frac{\pi}{2} \leq y \leq \frac{\pi}{2}$.

Algebra and Functions; Properties of Functions, Inverse Trigonometric

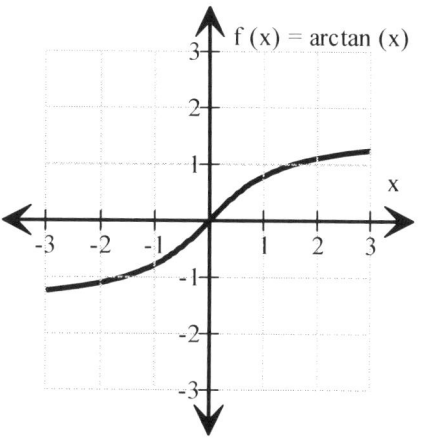

   (C)   II and III only

   (D)   I and III only

   (E)   I, II, and III

48. A bond pays annual coupons of $600 at the end of each year for 10 years and returns the original $10,000 investment at the end of the 10 years. What is the present value of the bond, with the cash flows discounted using an interest rate of 3%?

(A) $12,342

(B) $12,496

*(C) $12,559

The present value of the $600 annual coupon payments is the sum of a finite geometric series. The present value of the $10,000 payable in 10 years is $10,000 divided by 10 years of compound interest using the rate of inflation.

$$Present\ value\ of\ coupons = \frac{\$600}{1.03} + \frac{\$600}{1.03^2} + \frac{\$600}{1.03^3} + \cdots + \frac{\$600}{1.03^{10}}$$

$$= \frac{1}{1.03}\left[\$600 + \frac{\$600}{1.03} + \frac{\$600}{1.03^2} + \frac{\$600}{1.03^3} + \cdots + \frac{\$600}{1.03^9}\right] = \frac{S_n}{1.03}$$

$a_1 = \$600$

$r = \dfrac{1}{1.03}$

$$S_n = \frac{a_1(1-r^n)}{1-r} = \frac{\$600\left(1-\left(\frac{1}{1.03}\right)^{10}\right)}{1-\left(\frac{1}{1.03}\right)} = \frac{\$600\,(0.255906)}{0.029126} = \$5,271.67$$

$$\frac{S_n}{1.03} = \frac{\$5,271.67}{1.03} = \$5,118.12$$

$$Present\ value\ of\ \$10,000 = \frac{\$10,000}{1.03^{10}} = \frac{\$10,000}{1.343916} = \$7,440.94$$

$Present\ value\ of\ bond = \$5,118.12 + \$7,440.94 \approx \$12,559$

Number and Operations; Series

(D) $12,713

(E) $12,871

49.

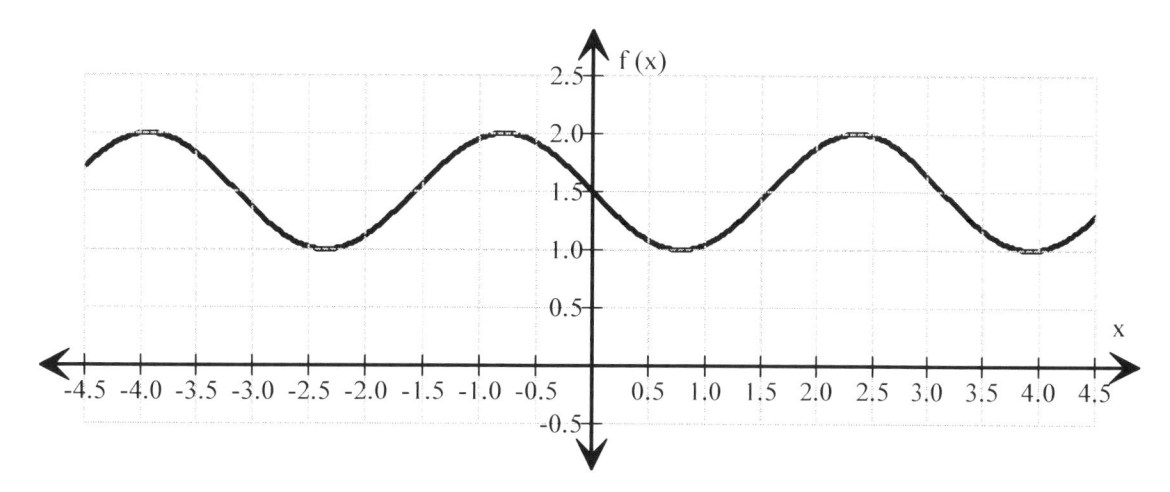

The graph above could represent which of the following functions?

(A) $f(x) = 2 \sin(0.5x + 2\pi) - 1.5$

(B) $f(x) = 2 \sin(0.5x + \pi) + 1.5$

(C) $f(x) = 2 \sin(0.5x + 2\pi) + 1.5$

(D) $f(x) = 0.5 \sin(2x + \pi) - 1.5$

`*(E) $f(x) = 0.5 \sin(2x + \pi) + 1.5$

The general form of the sine function is:

$f(x) = A \sin(Bx + C) + D$

Amplitude A = the distance from the midpoint to the highest or lowest point of the function.

Period T = the distance between any two repeating points on the function.

The period of the sine function is: T = 2π / | B |

D = Vertical displacement, or 'y' shift, of the midpoint of the function above the x-axis

Phase shift = the amount of horizontal displacement of the function from its original position.

Phase shift = C / B

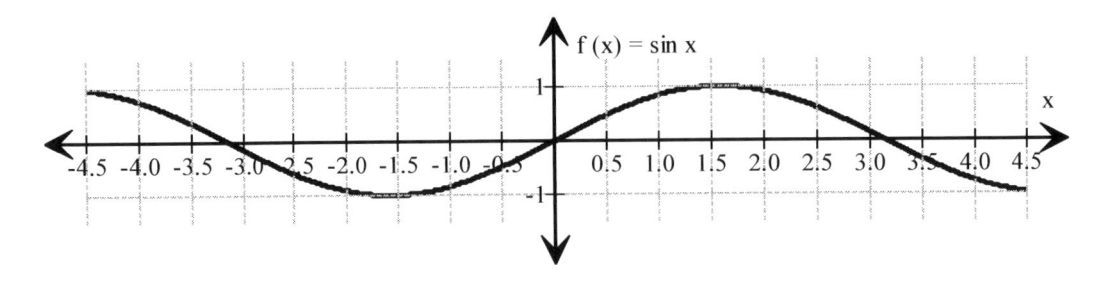

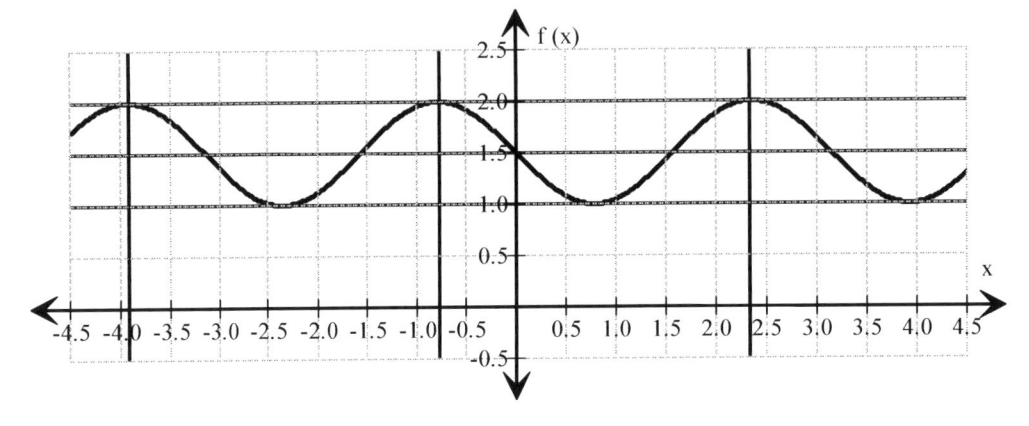

$A = |\,1.5 - 2.0\,| = |\,1.5 - 1.0\,| = 0.5$

$T = \pi = 2\pi / B$

$B = 2$

$D = 1.5$

$C / B = \pi / 2$

$C / 2 = \pi / 2$

$C = \pi$

$f(x) = A \sin(Bx + C) + D = 0.5 \sin(2x + \pi) + 1.5$

x	0.0	0.8	1.6	2.4
$f(x)$	1.5	1.0	1.0	2.0

Algebra and Functions; Properties of Functions, Trigonometric

50. What is the equation of the hyperbola having foci at (1, 8.4) and (1, −4.4) and vertices (1, 7) and (1, −3)?

(A) $(1/5)(x-2)^2 - (1/4)(y-1)^2 = 1$

(B) $(1/5)(y-2)^2 - (1/16)(x-1)^2 = 1$

*(C) $(1/25)(y-2)^2 - (1/4)(x-1)^2 = 1$

The general equation for a hyperbola opening left and right is:

$(1/a^2)(x-h)^2 - (1/b^2)(y-k)^2 = 1$

The general equation for a hyperbola opening up and down is:

$(1/a^2)(y-k)^2 - (1/b^2)(x-h)^2 = 1$

The center (h, k) is $(1 + |1 - 1|/2, -3 + |-3 - 7|/2) = (1, 2)$

The distance from the center to a focus point = $c = 6.4$

a = The distance from the center to the vertex (1, 7) is 5.

$a^2 = 5^2 = 25$

$b^2 = (6.4^2 - 5^2)^{0.5} = 4$

The hyperbola opens up and down because the axis connecting the vertices is vertical.

$1/a^2 = 1/5$

$1/b^2 = 1/4$

$(1/5)(y-2)^2 - (1/4)(x-1)^2 = 1$

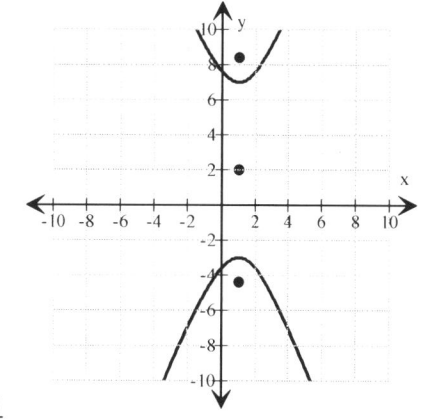

x	0	2	5
y	7.6 or −3.6	7.6 or −3.6	13.2 or −9.2

Geometry and Measurement; Coordinate, Hyperbolas

(D) $(1/25)(y-2)^2 - (1/16)(x-1)^2 = 1$

(E) $(1/25)(y-2)^2 + (1/16)(x-1)^2 = 1$

Math Level 2 Exam 2

Math Level 2 Exam 2

Time – 1 Hour
50 Questions

Directions: For this section, solve each problem and decide which answer is the best. Numerical answers may be rounded. A scientific or graphing calculator will be necessary for some questions. Angles are in degrees. Unless specified otherwise, the domain of any function is the set of all real numbers. Figures are not necessarily drawn to scale.

Reference Information

Volume of a right circular cone with radius r and height h: $V = \frac{1}{3} \pi r^2 h$

Volume of a sphere with radius r: $V = \frac{4}{3} \pi r^3$

Volume of a pyramid with base area B and height h: $V = \frac{1}{3} B h$

Surface area of a sphere with radius r: $S = 4 \pi r^2$

---

1. Evaluate *arcsin* 0.25882.

    (A) 0.3°

    (B) 7.4°

    (C) 8.6°

    (D) 14.8°

    (E) 15.0°

2. If ln (x) = 2.3, then ln (2x) = ?

    (A)   1.7
    (B)   2.3
    (C)   3.0
    (D)   4.6
    (E)   5.3

3. What is the radian measure of the angle whose degree measure is 45°?

    (A) $\dfrac{\pi}{8}$

    (B) $\dfrac{\pi}{4}$

    (C) $\dfrac{3\pi}{8}$

    (D) $\dfrac{\pi}{2}$

    (E) $\dfrac{5\pi}{8}$

4.   0, 1, 4, 9, 16, …

    Which of the following functions could describe the above sequence, given that $f(0) = 0$?

    (A) $f(x) = (x)(f(x-1))$
    (B) $f(x) = (x-1)(f(x-1))$
    (C) $f(x) = (x+1)(f(x-1)) - 1$
    (D) $f(x) = f(x-1) + 2x - 1$
    (E) $f(x) = (2x)(f(x-1)) - 2$

5. Which of the following could be the coordinates of the center of a circle tangent to the x-axis and the y-axis?

   (A) $(-2, 3)$

   (B) $(1, 3)$

   (C) $(2, -3)$

   (D) $(-4, -1)$

   (E) $(-1, 1)$

6. A line has parametric equations $y = 2 + t$ and $x = 3 - 0.2t$, where t is the parameter. What is the slope of the line?

   (A) -5

   (B) -0.2

   (C) 0.2

   (D) 1.5

   (E) 5

7. If $x - 3$ is a factor of $x^3 - 9x^2 + xt - 27$, then what is the value of ?

   (A) -27

   (B) -18

   (C) -6

   (D) 18

   (E) 27

   Algebra and Functions; Properties of Functions, Polynomial

8.

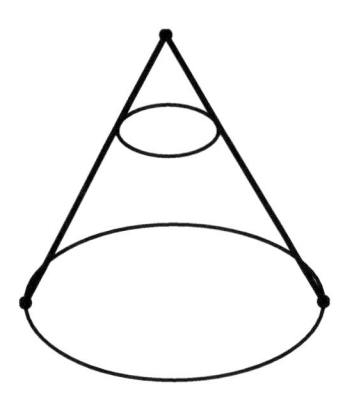

The right circular cone above is sliced horizontally forming two pieces. The height of the larger piece is twice the height of the smaller piece. What is the ratio of the volume of the larger piece to the volume of the smaller piece?

(A) 12

(B) 16

(C) 18

(D) 26

(E) 27

9. Evaluate $\dfrac{\arccos 0.5}{\arcsin 0.5}$.

(A) 0

(B) 1

(C) 2

(D) π

(E) undefined

10. Which of the following graphs is the graph of $y = csc(x)$?

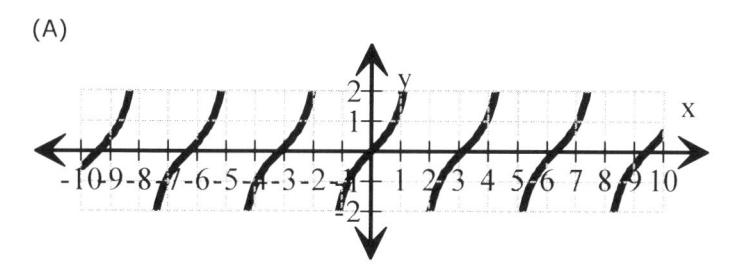

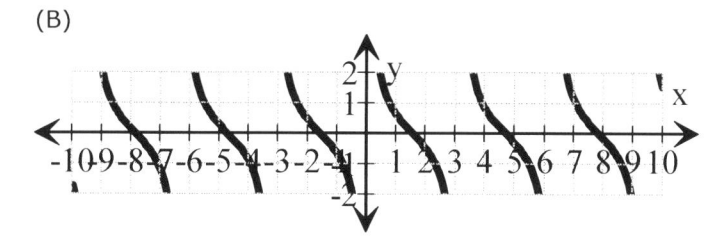

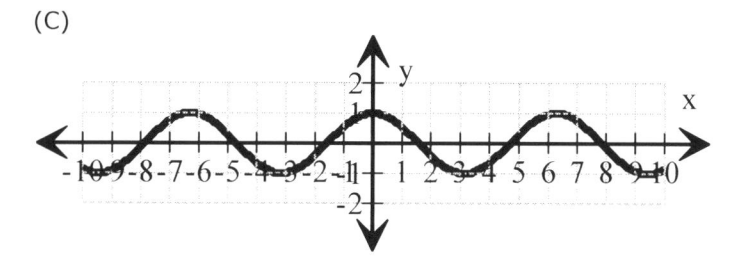

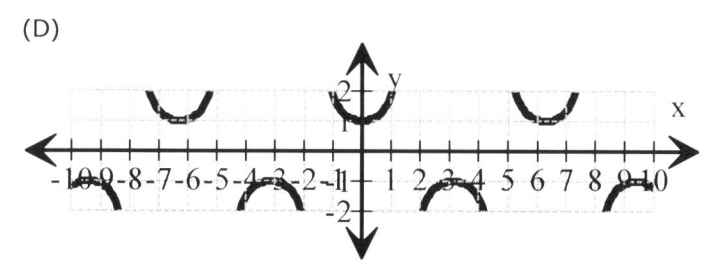

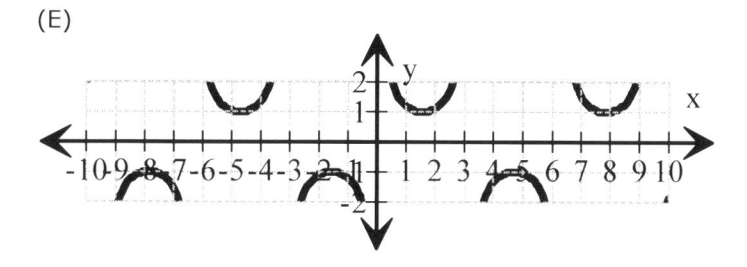

(A)

(B)

(C)

(D)

(E)

11. The vector (4, −7) is perpendicular to which of the following vectors?

   I.   (−7, −4)

   II. (−7, 4)

   III. (7, 4)

   (A)   I only

   (B)   II only

   (C)   III only

   (D)   I and II only

   (E)   I and III only

12. A right circular cylinder has radius 1 and height 3. If A, B, and C are three points on its surface, what is the maximum possible perimeter of triangle ABC ?

   (A) $4 + \sqrt{10}$

   (B) $4 + \sqrt{13}$

   (C) $5 + \sqrt{10}$

   (D) $2 + 2\sqrt{10}$

   (E) $5 + \sqrt{13}$

13. For some real number k, the first three terms of an arithmetic sequence are $t + 2$, $3t - 3$, and $2t + 4$. What is the numerical value of the fourth term?

   (A)   6

   (B)   8

   (C)   10

   (D)   12

   (E)   15

14. What is the range of the function defined by $f(x) = \dfrac{1}{x^2} - 3$?

   (A)   All real numbers greater than -3.

   (B)   All real numbers except -3.

   (C)   All real numbers greater than 0.

   (D)   All real numbers except 0.

   (E)   All real numbers.

15.

	Saturday	Sunday
Hamburgers	200	150
Smoothies	250	200

The table above shows the number of hamburgers and smoothies that were sold during a weekend promotion. The price of a hamburger was $2 and the price of the smoothie was $3. Which of the following matrix representations gives the total revenue, in dollars, received from the sale of the food for each of the two days?

   (A) $\begin{bmatrix} 200 & 150 \\ 250 & 200 \end{bmatrix} \begin{bmatrix} 2 & 3 \end{bmatrix}$

   (B) $\begin{bmatrix} 200 & 150 \\ 250 & 200 \end{bmatrix} \begin{bmatrix} 2 \\ 3 \end{bmatrix}$

   (C) $\begin{bmatrix} 2 & 3 \end{bmatrix} \begin{bmatrix} 200 & 150 \\ 250 & 200 \end{bmatrix}$

   (D) $\begin{bmatrix} 2 \\ 3 \end{bmatrix} \begin{bmatrix} 200 & 150 \\ 250 & 200 \end{bmatrix}$

   (E) $\begin{bmatrix} 2 \\ 3 \end{bmatrix} \begin{bmatrix} 200 & 250 \\ 150 & 200 \end{bmatrix}$

16. Which of the following statements are accurate regarding inequalities?

   I. For any real numbers a and b, and if $a < b$ then $-a > -b$.

   II. If c is negative and $a < b$, then $ac < bc$ and $a/c < b/c$.

   III. If either a or b is negative (but not both), and if $a < b$, then $1/a < 1/b$.

   (A)   I only

   (B)   I and II only

   (C)   I and III only

   (D)   II and III only

   (E)   I, II, and III

17. What is the population standard deviation of 4, 7, and 7 ?

   (A)   1.0

   (B)   1.4

   (C)   1.5

   (D)   1.7

   (E)   3.0

18. Evaluate $\tan(\arctan x)$ in radians.

   (A)   0

   (B)   1

   (C)   π

   (D)   x

   (E)   undefined

19. Using the rules of a classic board game, what are the odds that on a player's turn, the player will be "caught speeding" for throwing the pair of standard six-sided dice three times, and getting three doubles in a row?

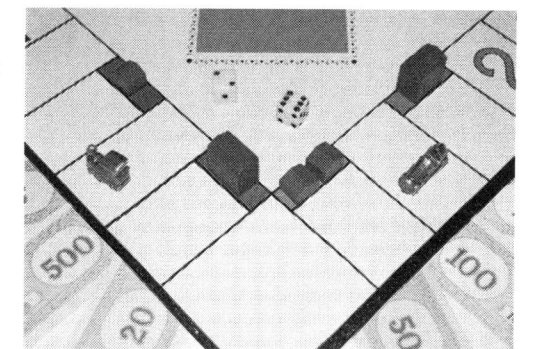

   (A) $1/1{,}296$

   (B) $1/324$

   (C) $1/216$

   (D) $1/108$

   (E) $1/36$

20. What is the range of the function defined by

$$f(x) = \begin{cases} x^3, & |x > -3 \\ 3x - 2, & |x \leq 3 \end{cases}$$

   (A) $-27 < y \leq -11$

   (B) $y \leq -11$

   (C) $y > -27$

   (D) $y \leq -11$

   (E) All real numbers.

21. Which of the following expressions is equivalent to $(\tan(2\alpha))(\tan^2 \alpha)$?

   (A) $\tan(2\alpha) - \tan \alpha$

   (B) $\tan \alpha + 2 \tan \alpha$

   (C) $\tan \alpha - 2 \tan \alpha$

   (D) $\tan(2\alpha) + 2 \tan \alpha$

   (E) $\tan(2\alpha) - 2 \tan \alpha$

22. A sequence is recursively defined by $a_n = a_{n-1} - 2a_{n-2}$, for $n > 2$. If $a_1 = 0$ and $a_2 = 1$, what is the value of a_{11}?

(A) -17

(B) -11

(C) -3

(D) 5

(E) 7

23. If, in a right triangle, $\sin\theta = x/4$, where $0 < \theta < \pi/2$ and $0 < x < 4$, then $\cos\theta =$

(A) $(\sqrt{16 - x^2})/4$

(B) $(\sqrt{16 + x^2})/4$

(C) $(\sqrt{4 - x^2})/16$

(D) $(\sqrt{4 + x^2})/16$

(E) $(\sqrt{4 - x^2})/8$

24. What is the period of the graph of $= 3 \tan(2\pi x - 5)$?

(A) $1/(3\pi)$

(B) $1/(2\pi)$

(C) $1/5$

(D) $1/2$

(E) $\pi/2$

25.
$$19\%, 8\%, -10\%, 14\%, -12\%, 0\%, 9\%, 25\%, 16\%, -1\%$$

If the percentiles are in the range 0 to 100, exclusive, what is the interquartile range for the rates of return shown above?

(A) 16%

(B) 17%

(C) 20%

(D) 29%

(E) 37%

26. If $f(x) = \sqrt[3]{2x^3 - 1}$, what is $f^{-1}(0.4)$?

(A) 0.65

(B) 0.81

(C) 0.91

(D) 0.95

(E) 1.02

McCaulay's Practice Exams for the SAT* Subject Test in Mathematics Level 2

27.

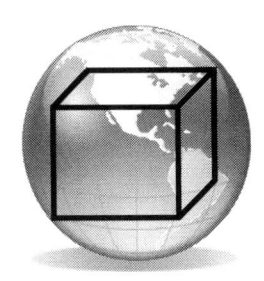

The cube above is inscribed in a sphere. The cube has a surface area of 50 square inches. What is the volume of the sphere?

(A)   26.2

(B)   35.6

(C)   36.8

(D)   37.9

(E)   65.4

28. As of September 2005, a company's revenue was $6.4 million. Assuming a growth rate of 5% per year, the company's revenue, in millions, for n years after 2010 can be modeled by the equation $R = \$6.4\,(1.05)^n$. According to the model, what was the company's revenue growth from September 2009 to September 2010?

(A) $389,000

(B) $397,000

(C) $408,000

(D) $417,000

(E) $429,000

Page 106

29. What is the measure of one of the larger angles of a parallelogram in the xy-plane that has vertices with coordinates (6, −3), (3, −7), (7, 7), and (4, −3) ?

(A) 106.0°

(B) 106.2°

(C) 106.7°

(D) 106.9°

(E) 108.0°

30. The population of the United States is projected to grow exponentially to 400 million in the year 2050, a 30% increase from the population in the year 2010. What is the projected population in the year 2040, in millions?

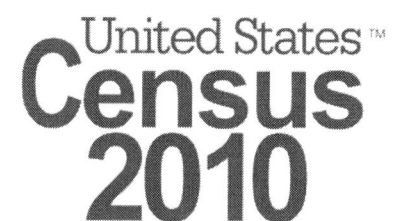

(A) 368

(B) 370

(C) 374

(D) 375

(E) 377

31. Find the exponential function that best fits the data below.

x	0	1	2	3	4	5
y	1.7	0.5	0.16	0.05	0.016	0.005

(A) $y = (1.65)(0.31^x)$

(B) $y = (1.68)(0.31^x)$

(C) $y = (1.65)(0.32^x)$

(D) $y = (1.68)(0.32^x)$

(E) $y = (1.70)(0.30^x)$

32. After eliminating the parameter, what is the rectangular equation whose graph represents the parametric curve $x = t^2/2$ and $= t^2 + 1$?

(A) $y = 4x^2 + 1$ for all real numbers

(B) $y = 4x^2 + 1$ for $x \geq 0$

(C) $y = 2x + 1$ for all real numbers

(D) $y = 2x + 1$ for $x \geq 0$

(E) $y = (x-1)^2/2$ for $x \geq 1$

33.

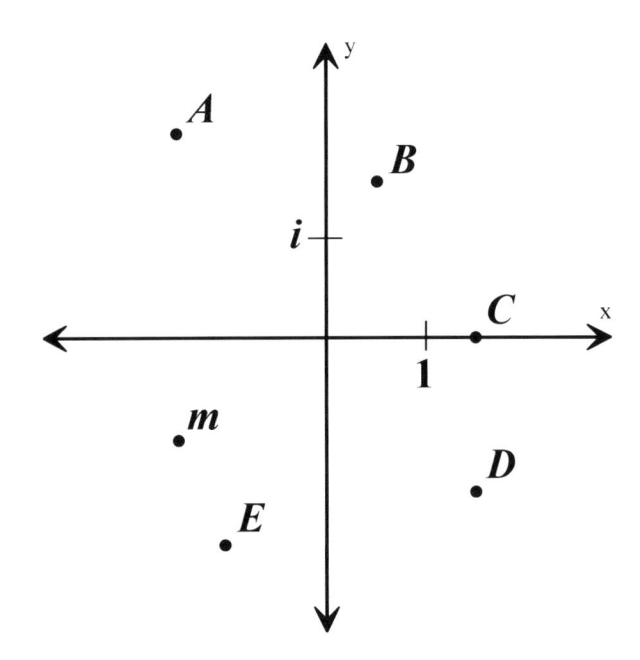

If m is the complex number shown in the figure above, which of the following could be im ?

(A) A

(B) B

(C) C

(D) D

(E) E

34. Which of the following graphs represents the greatest integer function, $y = [x]$, where $[x]$ is the greatest integer less than or equal to x ?

(A)

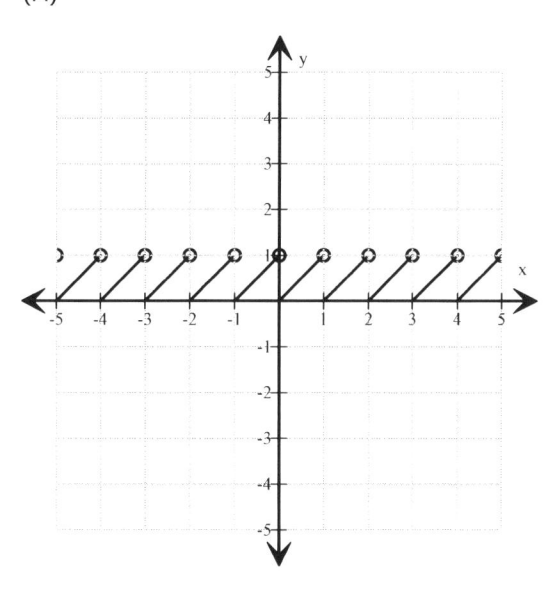

(B)

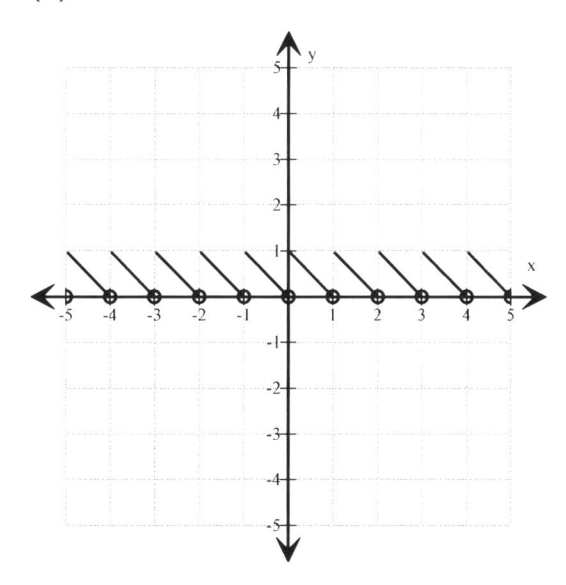

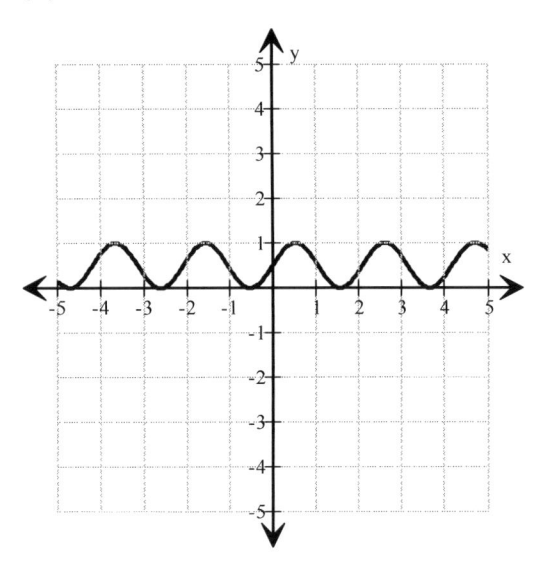

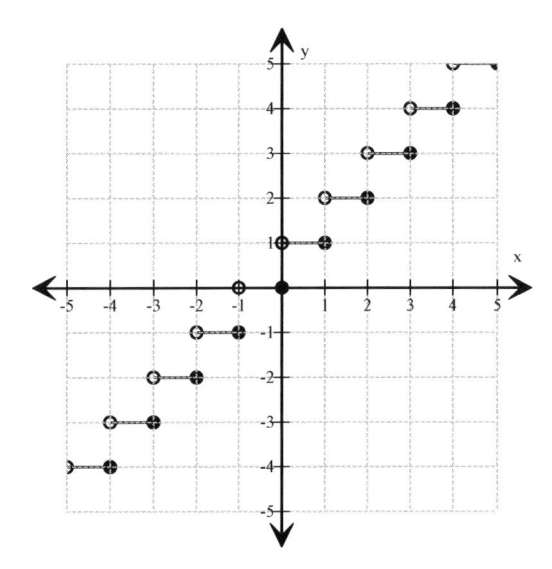

(C)

(D)

(E)

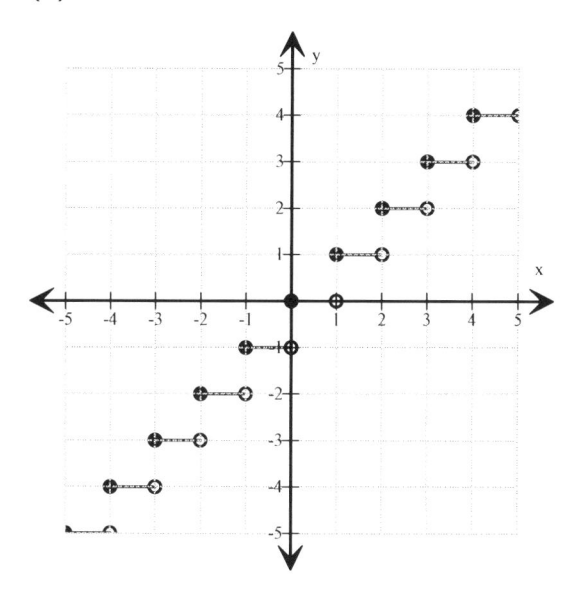

35. Express the point with polar coordinates $(2, \frac{3\pi}{10})$ in terms of rectangular coordinates.

    (A)   (0.6, 0.8)

    (B)   (1.0, 1.7)

    (C)   (1.2, 1.6)

    (D)   (1.4, 1.6)

    (E)   (1.6, 1.2)

36. The population of a colony of bacteria increases exponentially. There are 7,000 bacteria at the start. The formula $N = N_0\, e^{0.06\,t}$ gives the population after t hours. To the nearest hour, how many hours will it take for the population to reach 10,000?

(A) 5

(B) 6

(C) 7

(D) 8

(E) 9

37. A lottery pays $1 per year for an infinite number of years. What is the present value of the payment stream if the cash flows are discounted at an interest rate of 8% per year and the first of the annual $1 payments starts immediately?

(A) $8.64

(B) $10.00

(C) $12.50

(D) $13.50

(E) $25.00

38. If $0 < \theta < \pi/2$ and $\cos \theta = x$, then $\tan \theta =$

  (A) $x/(\sqrt{1 - x^2})$

  (B) $x/(1 - x^2)$

  (C) $1/x$

  (D) $(\sqrt{1 - x^2})/x$

  (E) $(1 - x^2)/x$

39. Which of the following graphs represents the curve determined by the parametric equations $x = 2t - 3$ and $y = t^2 - 5$ on the interval $0 \le t \le 3$?

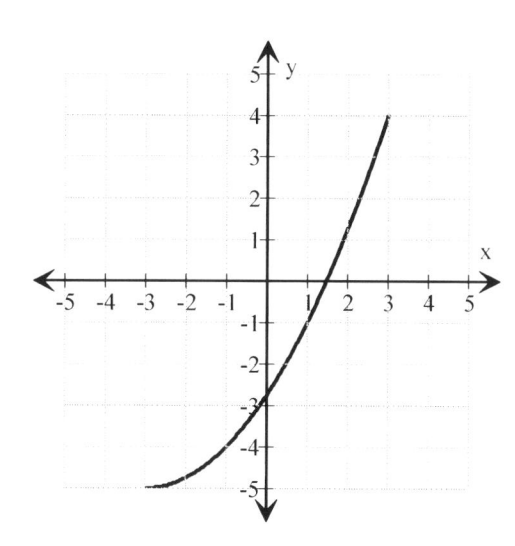

  (A)

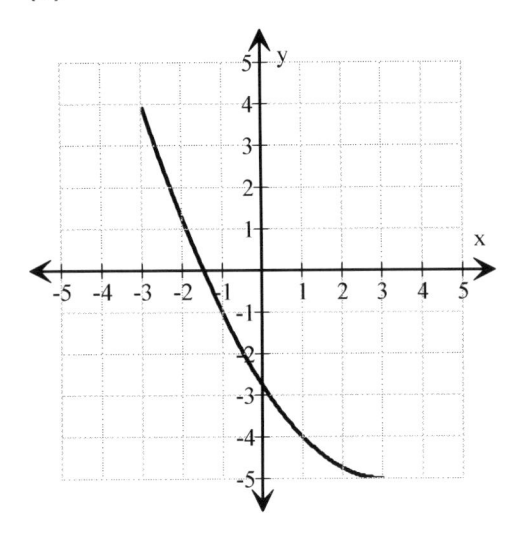

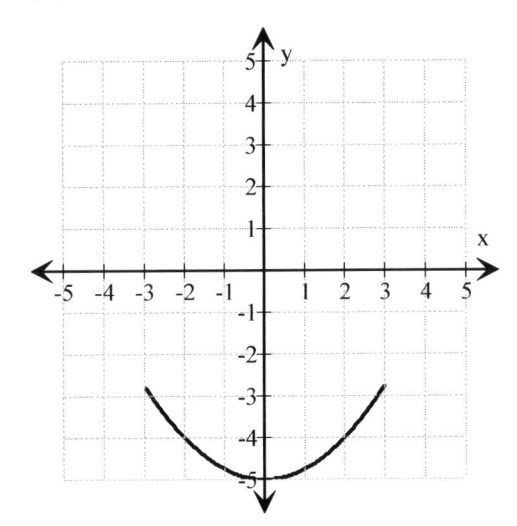

(B)

(C)

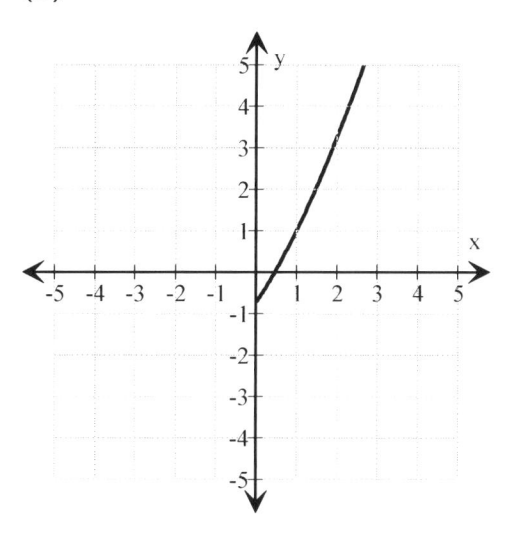

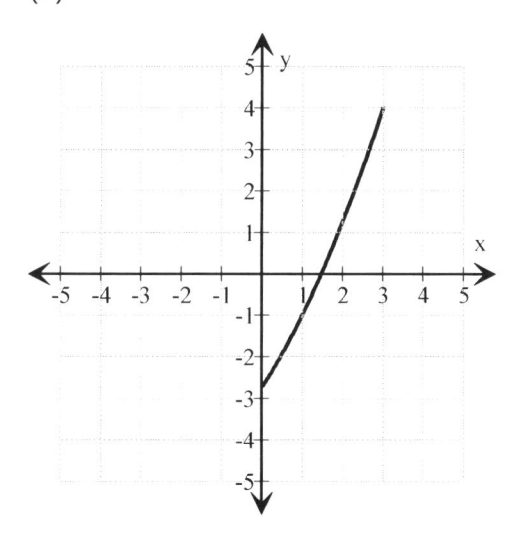

(D)

(E)

40. Using the quadratic regression model, which parabola best fits the data for the number of Hummers sold in the U.S., in thousands, and the number of years since 2000?

Years since 2000	0	1	2	3	4	5	6	7	8	9	10
Number Sold (thousands)	1	1	20	35	29	57	72	56	27	8	2

(A) $y = -3x^2 + 26x - 15$

(B) $y = -2x^2 + 24x - 13$

(C) $y = 2x^2 - 26x - 13$

(D) $y = 2x^2 - 24x + 13$

(E) $y = 3x^2 - 26x + 15$

41. What is the length of the major axis of the ellipse whose equation is $48x^2 + 16y^2 = 112$

(A) 1.7

(B) 2.6

(C) 3.5

(D) 5.3

(E) 6.9

42. In a savings account that earns 4% compounded annually, how many years would it take for a customer to triple the amount initially deposited?

(A) 18

(B) 21

(C) 28

(D) 32

(E) 36

43. A standard 52-card deck of playing cards has four aces. What is the probability of being dealt a five-card hand that includes four aces?

   (A) $1/270{,}725$

   (B) $1/54{,}145$

   (C) $1/20{,}825$

   (D) $1/16{,}660$

   (E) $1/4{,}165$

44. Which of the following graphs is the best parametric representation of the equations $x = 3\sec(t)$ and $y = \tan(t)$?

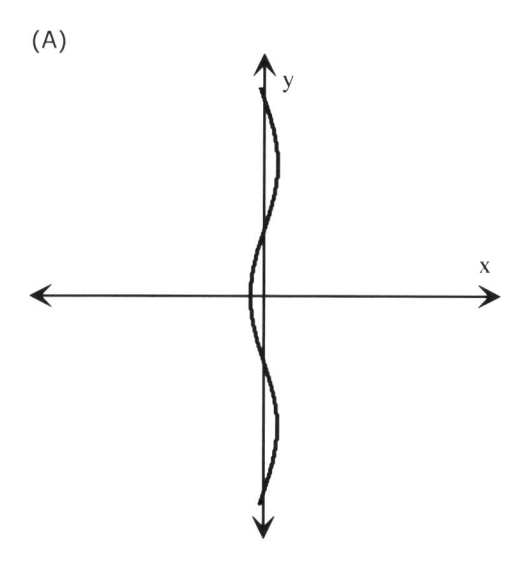

   (A)

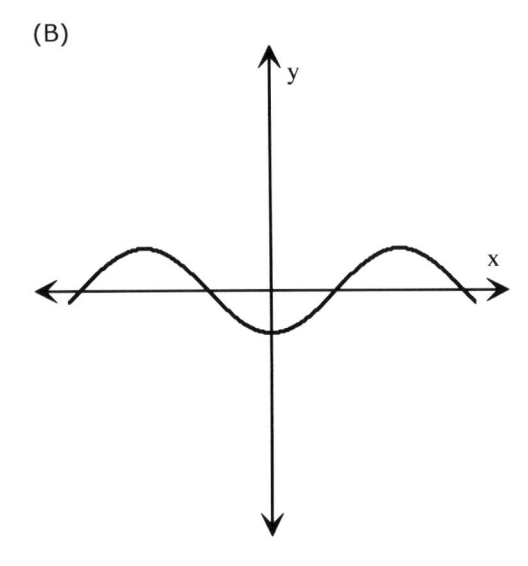

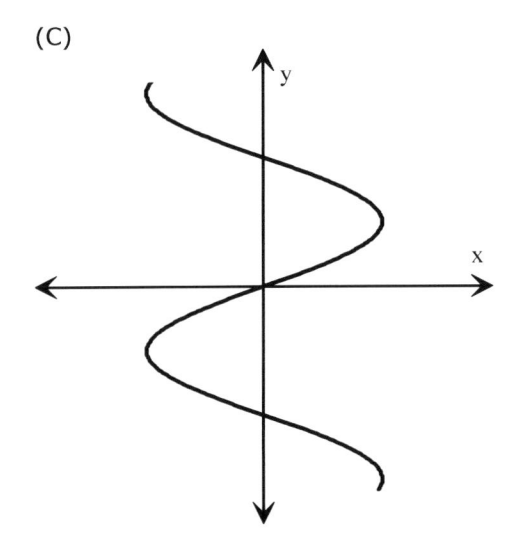

(B)

(C)

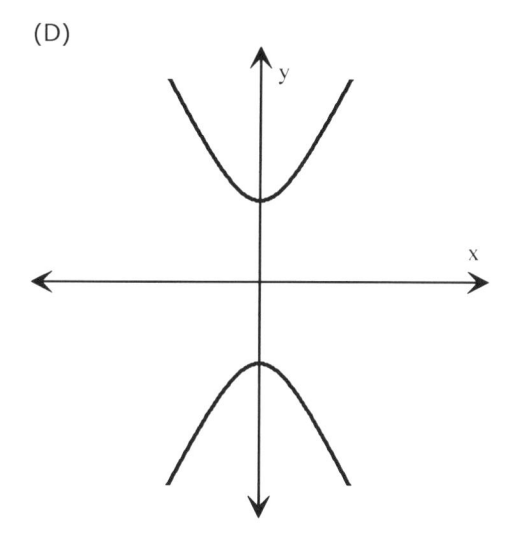

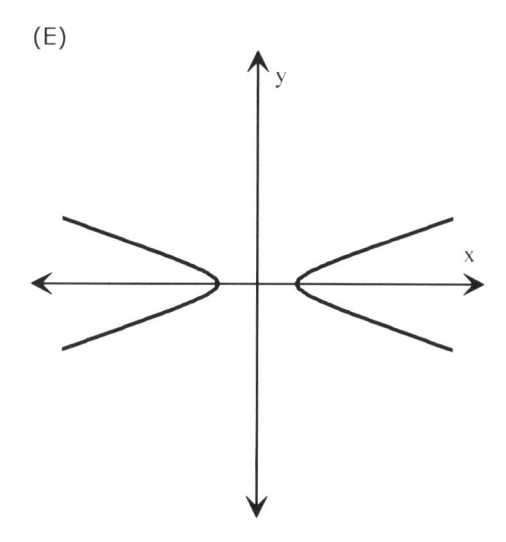

(D)

(E)

45.

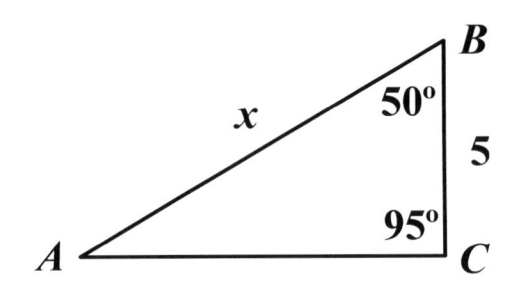

Note: Figure not drawn to scale.

In the figure above, $BC = 5$, $\angle B = 50°$, and $\angle C = 95°$. What is the value of x ?

(A)   6.53

(B)   6.68

(C)   8.68

(D)   8.72

(E)   8.75

46. Suppose the graph of $f(x) = x^2 - 2x + 2$ is translated 2 units right and 1 unit down. If the resulting graph represents $g(x)$, what is the value of $g(1.5)$?

(A)   2.25

(B)   3.25

(C)   4.25

(D)   6.25

(E)   8.25

47. Which of the following functions represents a sine function with a maximum of 4, a minimum of −2, a period of $\pi/4$, and no horizontal shift?

(A) $f(x) = 3\sin(8x) + 1$

(B) $f(x) = 3\sin(4x) - 1$

(C) $f(x) = 3\sin(4x) + 2$

(D) $f(x) = 6\sin(3x) - 1$

(E) $f(x) = 6\sin(8x) + 1$

48. A new employee starts a job with an annual salary of $100,000. The first pay increase will be granted after one year. Each annual pay increase is 5% of the prior year's salary. How much will the employee earn in total salaries over a 30 year career?

(A) $3,021,856

(B) $5,175,000

(C) $6,232,271

(D) $6,643,885

(E) $6,976,079

49. The polyhedron shown below consists of a cube with sides of length S with an identical set of eight cubes attached to each of the six faces of the cube, for a total of 49 cubes. A total of 174 square pyramids with the lengths of the sides of the pyramid bases equal to the lengths of the sides of the cubes will be attached to 174 of the 198 exposed cube faces to make a 720-faced polyhedron. The height of each of the 174 square pyramids equals $(s^4 - 49s)/58$. What is the volume of the polyhedron formed by the 49 cubes and the 174 square pyramids?

(A) $78 s^3$

(B) $49 s^3 + 58 s^4$

(C) $49 s^3 + s^6$

(D) $49 s^3 + 174 s^2$

(E) s^6

50.

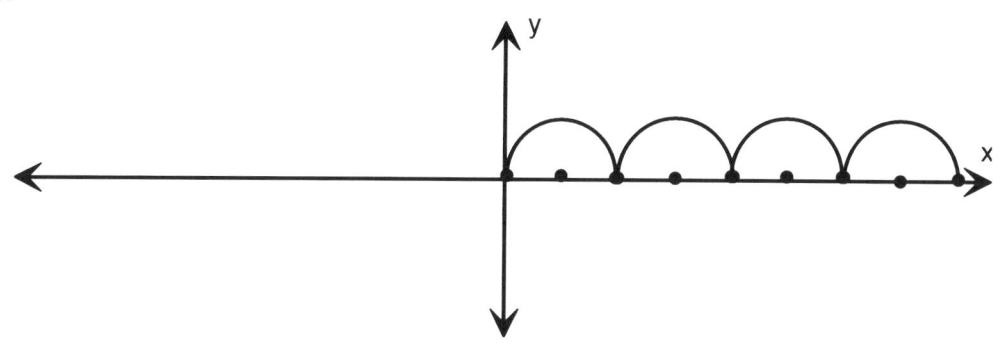

Which of the following parametric equations is represented by the above graph?

(A) $x = \alpha(t - \sin(t))$ and $y = \alpha(t - \cos(t))$

(B) $x = \alpha(1 + \sin(t))$ and $y = \alpha(1 + \cos(t))$

(C) $x = \alpha(t + \sin(t))$ and $y = \alpha(t + \cos(t))$

(D) $x = \alpha(1 - \sin(t))$ and $y = \alpha(t - \cos(t))$

(E) $x = \alpha(t - \sin(t))$ and $y = \alpha(1 - \cos(t))$

Math Level 2 Exam 2 Answers

1	E	21	E	41	D
2	C	22	B	42	C
3	B	23	A	43	B
4	D	24	D	44	E
5	E	25	C	45	C
6	A	26	B	46	A
7	E	27	E	47	A
8	D	28	A	48	D
9	C	29	C	49	E
10	E	30	D	50	E
11	E	31	A		
12	E	32	D		
13	E	33	D		
14	A	34	E		
15	C	35	C		
16	C	36	B		
17	B	37	D		
18	D	38	D		
19	C	39	A		
20	E	40	B		

Math Level 2 Exam 2 Explanations

1. Evaluate *arcsin* 0.25882.

    (A)   0.3°

    (B)   7.4°

    (C)   8.6°

    (D)   14.8°

    *(E)   15.0°

    sin θ = 0.25882 when θ = 15°

    arcsin 0.25882 = 15°

    Algebra and Functions; Properties of Functions, Inverse Trigonometric

2. If ln (x) = 2.3, then ln $(2x)$ = ?

   (A)   1.7

   (B)   2.3

   *(C) 3.0

   ln $(2x)$ = ln (2) + ln (x) = 0.69 + 2.3 ≈ 3

   x = 10

   ln (10) = 2.3

   ln (20) = 3.0

   Algebra and Functions; Properties of Functions, Logarithmic

   (D)   4.6

   (E)   5.3

3. What is the radian measure of the angle whose degree measure is 45°?

(A) $\dfrac{\pi}{8}$

*(B) $\dfrac{\pi}{4}$

$$1° = \left(\dfrac{\pi}{180}\right) radians$$

$$45° = \left(\dfrac{45\pi}{180}\right) radians = \left(\dfrac{\pi}{4}\right) radians$$

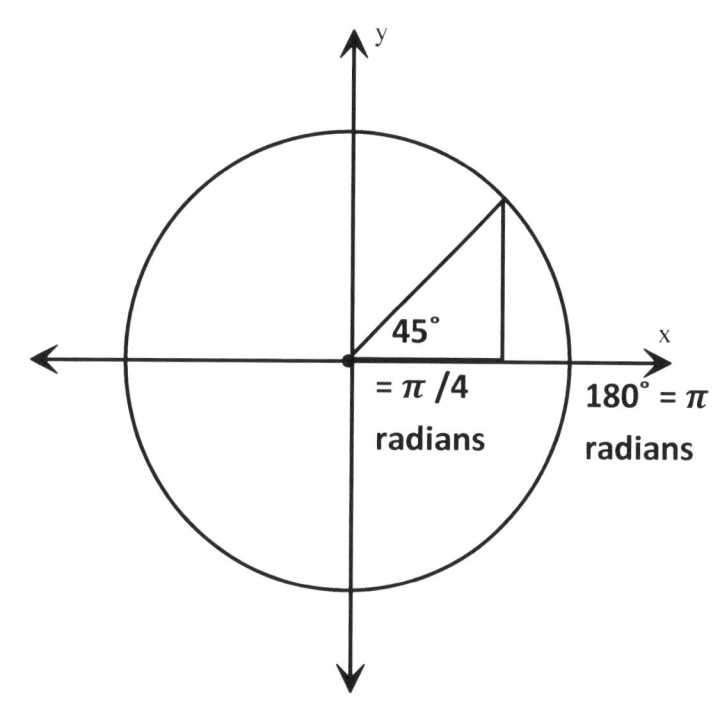

$$\sin 45° = 0.707;\ \sin\left(\dfrac{\pi}{4}\right) = 0.707$$

Geometry and Measurement; Trigonometry, Radian Measure

(C) $\dfrac{3\pi}{8}$

(D) $\dfrac{\pi}{2}$

(E) $\dfrac{5\pi}{8}$

4.   0, 1, 4, 9, 16, ...

Which of the following functions could describe the above sequence, given that $f(0) = 0$?

(A) $f(x) = (x)\,(f(x-1))$

(B) $f(x) = (x-1)\,(f(x-1))$

(C) $f(x) = (x+1)\,(f(x-1)) - 1$

*(D) $f(x) = f(x-1) + 2x - 1$

x	$x-1$	$f(x-1)$	$2x$	$f(x-1) + 2x - 1$
1	0	0	2	1
2	1	1	4	4
3	2	4	6	9
4	3	9	8	16
5	4	16	10	25

This is the recursive definition for x^2.

Algebra and Functions; Properties of Functions, Recursive

(E) $f(x) = (2x)\,(f(x-1)) - 2$

5. Which of the following could be the coordinates of the center of a circle tangent to the x-axis and the y-axis?

(A) $(-2, 3)$

(B) $(1, 3)$

(C) $(2, -3)$

(D) $(-4, -1)$

*(E) $(-1, 1)$

For a circle to be tangent to both the x-axis and the y-axis, the absolute values of the x-coordinate and the y-coordinate need to be equal.

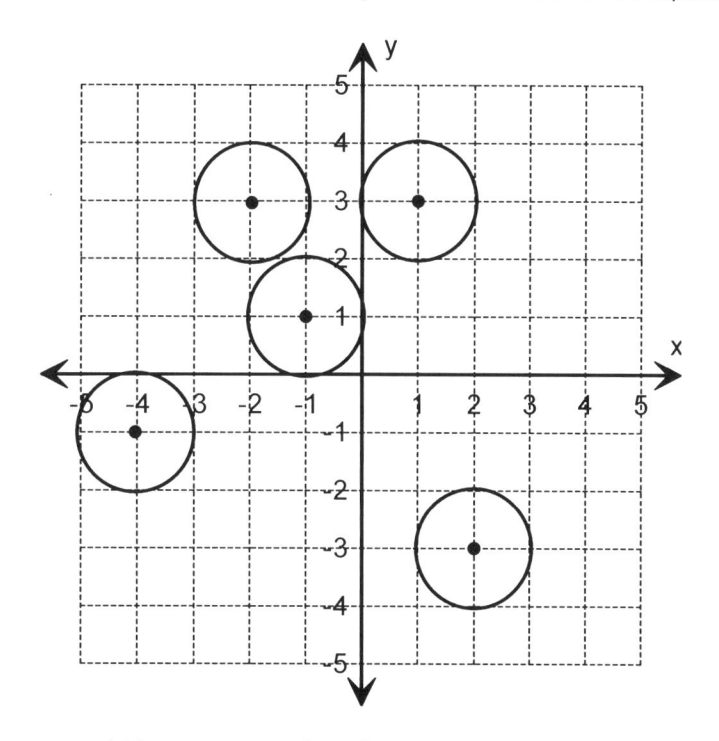

Geometry and Measurement, Coordinate, Circles

6. A line has parametric equations $y = 2 + t$ and $x = 3 - 0.2\,t$, where t is the parameter. What is the slope of the line?

*(A)   −5

$y = 2 + t$

$t = y - 2$

$x = 3 - 0.2\,t$

$0.2\,t = 3 - x$

$t = 15 - 5\,x$

$y - 2 = 15 - 5\,x$

$y = -5\,x + 17$

$y = m\,x + b$

$m = -5$

When $t = 0$, $(x, y) = (3, 2)$

When $t = 5$, $(x, y) = (2, 7)$

$m = (7 - 2) / (2 - 3) = -5$

Algebra and Functions; Properties of Functions, Parametric

(B)   −0.2

(C)   0.2

(D)   1.5

(E)   5

7. If $x - 3$ is a factor of $x^3 - 9x^2 + xt - 27$, then what is the value of ?

(A) -27

(B) -18

(C) -6

(D) 18

*(E) 27

$(x - 3)^3 = x^3 - 9x^2 + 27x - 27$

$t = 27$

$x^3 - 9x^2 + tx - 27 = (x - 3)(x^2 + nx + 9)$

$= x^3 + nx^2 + 9x - 3x^2 - 3nx - 27$

$= x^3 + (n - 3)x^2 + (9 - 3n)x - 27$

$n - 3 = -9$

$n = -6$

$9 - 3n = 9 - 3(-6) = 9 + 18 = 27$

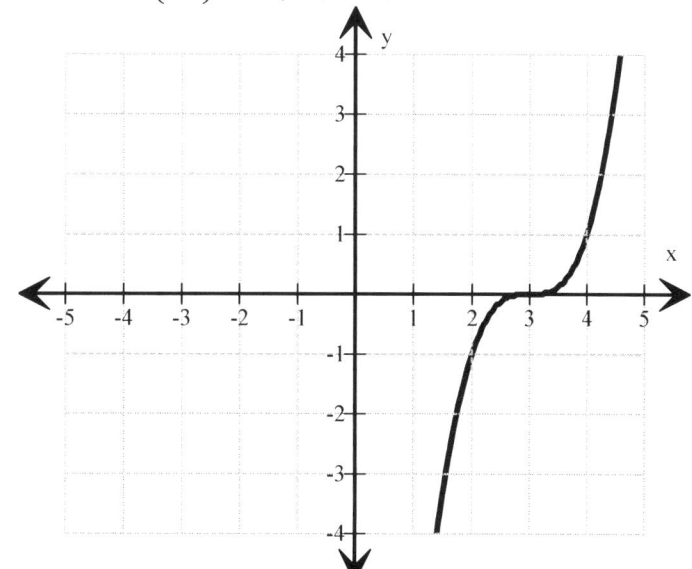

Algebra and Functions; Properties of Functions, Polynomial

8.

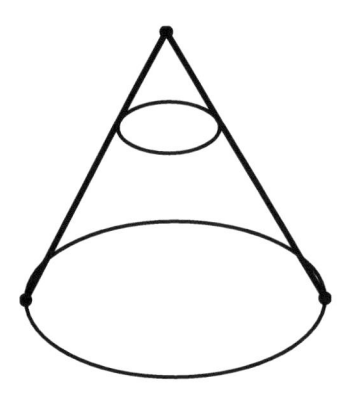

The right circular cone above is sliced horizontally forming two pieces. The height of the larger piece is twice the height of the smaller piece. What is the ratio of the volume of the larger piece to the volume of the smaller piece?

(A) 12

(B) 16

(C) 18

*(D) 26

The top piece is a cone whose height h is one-third the height of the original cone $3h$. Using the properties of similar right triangles, it can be determined that the radii of these two cones must be in the same ratio. If the top cone has radius r, the original cone has radius $3r$.

The volume of the top piece is equal to $\frac{1}{3} \pi r^2 h$. The volume of the bottom piece is equal to the volume of the original cone minus the volume of the top piece.

$$V = \frac{1}{3} \pi (3r)^2 (3h) - \frac{1}{3} \pi r^2 h = \frac{26}{3} \pi r^2 (2h)$$

$$[\frac{26}{3} \pi r^2 h] / [\frac{1}{3} \pi r^2 h] = 26$$

Geometry and Measurement; Surface Area and Volume of Solids, Cones

(E) 27

9. Evaluate $\dfrac{\arccos 0.5}{\arcsin 0.5}$.

(A)   0

(B)   1

*(C) 2

$\cos \theta = 0.5$ when $\theta = 60°$

$\arccos 0.5 = 60°$

$\sin \theta = 0.5$ when $\theta = 30°$

$\arcsin 0.5 = 30°$

$\dfrac{\arccos 0.5}{\arcsin 0.5} = \dfrac{60°}{30°} = 2$

$\dfrac{\arccos 0.5}{\arcsin 0.5} = \dfrac{1.0472}{0.5236} = 2$

Algebra and Functions; Properties of Functions, Inverse Trigonometric

(D) π

(E)   undefined

10. Which of the following graphs is the graph of $y = \csc(x)$?

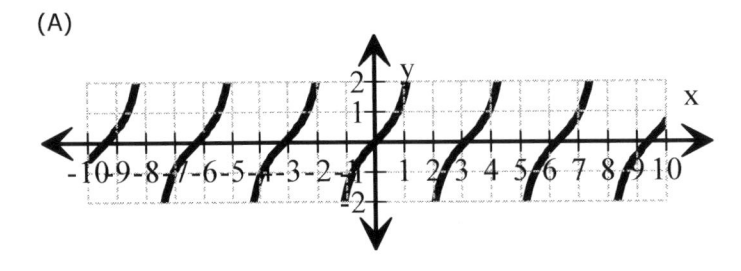

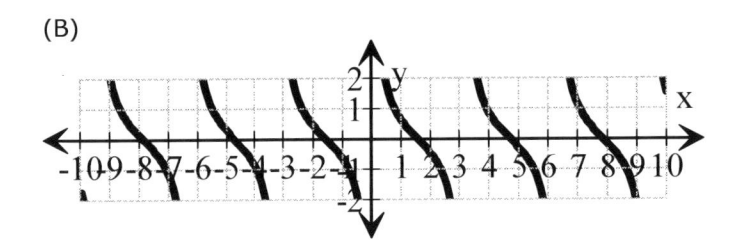

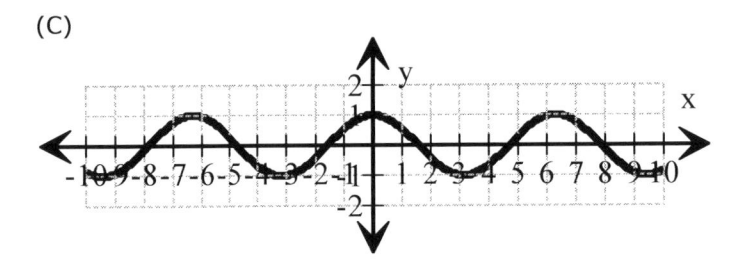

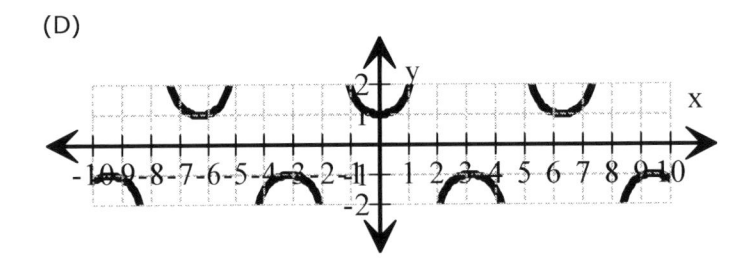

(A)

(B)

(C)

(D)

*(E)

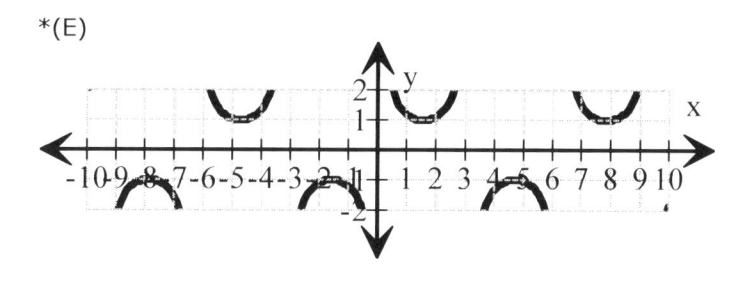

csc x = 1 / sin x

1 / sin(0) = 1 / 0 = undefined

1 / sin($\pi/2$) = 1 / 1 = 1

1 / sin(π) = 1 / 0 = undefined

1 / sin($3\pi/2$) = 1 / −1 = −1

Graph A is *tan(x)*

Graph B is 1 / *tan(x)* = *cot(x)*

Graph C is *cos(x)*

Graph D is 1 / *cos(x)* = *sec(x)*

Graph E is 1 / *sin(x)* = *csc(x)*

Algebra and Functions; Properties of Functions, Trigonometric

11. The vector (4, −7) is perpendicular to which of the following vectors?

    I.   (−7, −4)

    II. (−7, 4)

    III. (7, 4)

(A)   I only

(B)   II only

(C)   III only

(D)   I and II only

*(E)   I and III only

Vectors are perpendicular if the dot product = 0.

$v = (4, -7)$

$s = (-7, -4)$

$t = (-7, 4)$

$u = (7, 4)$

$v \bullet s = v_1 s_1 + v_2 s_2 = (4)(-7) + (-7)(-4) = -28 + 28 = 0$

$v \bullet t = v_1 t_1 + v_2 t_2 = (4)(-7) + (-7)(4) = -28 - 28 = -56 \neq 0$

$v \bullet u = v_1 u_1 + v_2 u_2 = (4)(7) + (-7)(4) = 28 - 28 = 0$

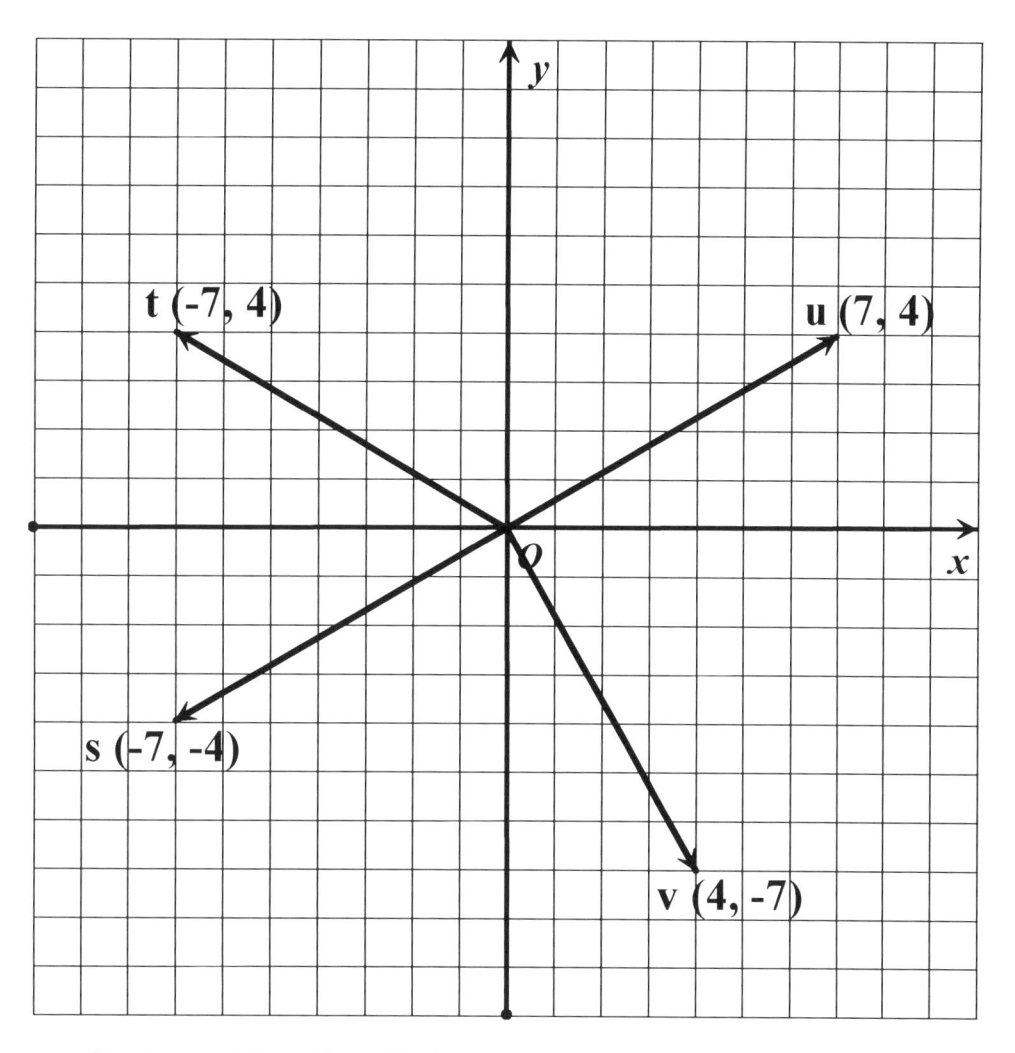

Number and Operations; Vectors

12. A right circular cylinder has radius 1 and height 3. If A, B, and C are three points on its surface, what is the maximum possible perimeter of triangle ABC ?

(A) $4 + \sqrt{10}$

(B) $4 + \sqrt{13}$

(C) $5 + \sqrt{10}$

(D) $2 + 2\sqrt{10}$

*(E) $5 + \sqrt{13}$

The triangle with the maximum perimeter has one side with a length equal to the diameter of base of the cylinder, one side with a length equal to the height of the cylinder, and one side equal to the diagonal that connects the other two sides.

Perimeter = $(2)(1) + 3 + \sqrt{2^2 + 3^2} = = 5 + \sqrt{13}$

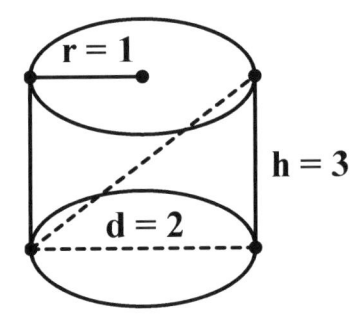

Geometry and Measurement; Coordinate, Lines

13. For some real number k, the first three terms of an arithmetic sequence are $t + 2$, $3t - 3$, and $2t + 4$. What is the numerical value of the fourth term?

   (A)   6

   (B)   8

   (C)   10

   (D)   12

   *(E)   15

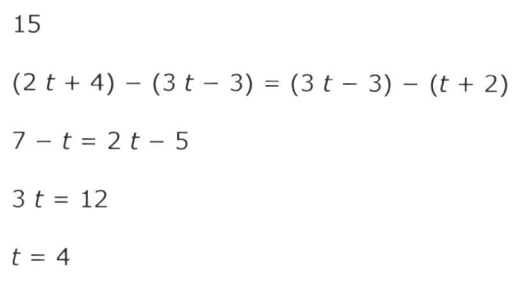

$(2t + 4) - (3t - 3) = (3t - 3) - (t + 2)$

$7 - t = 2t - 5$

$3t = 12$

$t = 4$

$t + 2, 3t - 3, 2t + 4, \ldots$

$6, 9, 12, \ldots$

The fourth term is $12 + 3 = 15$

Number and Operations; Sequences

14. What is the range of the function defined by $f(x) = \dfrac{1}{x^2} - 3$?

   *(A)   All real numbers greater than -3.

   $y = \dfrac{1}{x^2} - 3$

   $\dfrac{1}{x^2}$ is undefined when $x = 0$

   When $x \neq 0$; $\dfrac{1}{x^2} > 0$ and $\dfrac{1}{x^2} - 3 > -3$.

   The range of $f(x)$ is all real numbers greater than -3

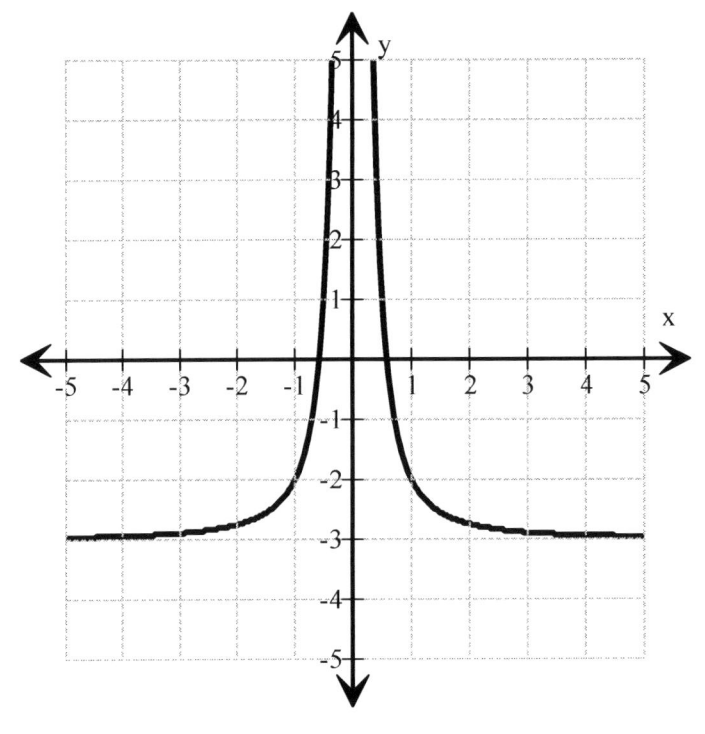

   Algebra and Functions; Properties of Functions, Rational

   (B)   All real numbers except -3.

   (C)   All real numbers greater than 0.

   (D)   All real numbers except 0.

   (E)   All real numbers.

15.

	Saturday	Sunday
Hamburgers	200	150
Smoothies	250	200

The table above shows the number of hamburgers and smoothies that were sold during a weekend promotion. The price of a hamburger was $2 and the price of the smoothie was $3. Which of the following matrix representations gives the total revenue, in dollars, received from the sale of the food for each of the two days?

(A) $\begin{bmatrix} 200 & 150 \\ 250 & 200 \end{bmatrix} \begin{bmatrix} 2 & 3 \end{bmatrix}$

(B) $\begin{bmatrix} 200 & 150 \\ 250 & 200 \end{bmatrix} \begin{bmatrix} 2 \\ 3 \end{bmatrix}$

*(C) $\begin{bmatrix} 2 & 3 \end{bmatrix} \begin{bmatrix} 200 & 150 \\ 250 & 200 \end{bmatrix}$

The revenue for Saturday was (200)($2) + (250)($3) = $1,150.

The revenue for Sunday was (150)($2) + (200)($3) = $900.

$\begin{bmatrix} 2 & 3 \end{bmatrix} \begin{bmatrix} 200 & 150 \\ 250 & 200 \end{bmatrix} = \begin{bmatrix} (2)(200)+(3)(250) & (2)(150)+(3)(200) \end{bmatrix}$

$= \begin{bmatrix} 1,150 & 900 \end{bmatrix}$

Number and Operations; Matrices

(D) $\begin{bmatrix} 2 \\ 3 \end{bmatrix} \begin{bmatrix} 200 & 150 \\ 250 & 200 \end{bmatrix}$

(E) $\begin{bmatrix} 2 \\ 3 \end{bmatrix} \begin{bmatrix} 200 & 250 \\ 150 & 200 \end{bmatrix}$

16. Which of the following statements are accurate regarding inequalities?

    I.   For any real numbers *a* and *b,* and if $a < b$ then $-a > -b$.

    II. If *c* is negative and $a < b$, then $ac < bc$ and $a/c < b/c$.

    III. If either *a* or *b* is negative (but not both), and if $a < b$, then $1/a < 1/b$.

(A)   I only

(B)   I and II only

*(C) I and III only

If *c* is positive and $a < b$, then $ac < bc$ and $a/c < b/c$

If *c* is negative and $a < b$, then $ac > bc$ and $a/c > b/c$

For any real numbers *a* and *b*:
   If $a < b$ then $-a > -b$
   If $a > b$ then $-a < -b$

For any non-zero real numbers *a* and *b* that are both positive or both negative:
   If $a < b$ then $1/a > 1/b$
   If $a > b$ then $1/a < 1/b$

If either a or b is negative (but not both) then:
   If $a < b$ then $1/a < 1/b$
   If $a > b$ then $1/a > 1/b$

Algebra and Functions; Inequalities

(D)   II and III only

(E)   I, II, and III

17. What is the population standard deviation of 4, 7, and 7 ?

   (A)   1.0

   *(B)   1.4

   Mean = (4 + 7 + 7) / 3 = 18 / 3 = 6

   The population standard deviation is found by adding the sums of the squares of the differences in the values and the mean, dividing by the number of values, and taking the square root.

   Population standard deviation

   $= [[(4 - 6)^2 + (7 - 6)^2 + (7 - 6)^2] / 3]^{0.5}$

   $= [[(-2)^2 + (1)^2 + (1)^2] / 3]^{0.5} = [(4 + 1 + 1) / 3]^{0.5} = (6 / 3)^{0.5}$

   $= \sqrt{2} \approx 1.4$

   The sample standard deviation is found by adding the sums of the squares of the differences in the values and the mean, dividing by the number of values minus 1, and taking the square root.

   Sample standard deviation

   $= [[(4 - 6)^2 + (7 - 6)^2 + (7 - 6)^2] / (3 - 1)]^{0.5}$

   $= [[(-2)^2 + (1)^2 + (1)^2] / 2]^{0.5} = [(4 + 1 + 1) / 2]^{0.5} = (6 / 2)^{0.5}$

   $= \sqrt{3} \approx 1.7$

   Data Analysis, Statistics, and Probability; Standard Deviation

   (C)   1.5

   (D)   1.7

   (E)   3.0

18. Evaluate *tan (arctan x)* in radians.

(A) 0

(B) 1

(C) π

*(D) x

The domain of $y = \tan^{-1} x$ is all real numbers.

$\tan(\tan^{-1} x) = x$

$\tan(\tan^{-1} \pi) = \tan(1.262627) = 3.14159$

$\tan(\tan^{-1} 0) = \tan(0) = 0$

Algebra and Functions; Properties of Functions, Inverse Trigonometric

(E) undefined

19. Using the rules of a classic board game, what are the odds that on a player's turn, the player will be "caught speeding" for throwing the pair of standard six-sided dice three times, and getting three doubles in a row?

(A) $1/1{,}296$

(B) $1/324$

*(C) $1/216$

There are (6) (6) = 36 combinations for each throw of the dice.

	1	2	3	4	5	6
1	**1, 1**	1, 2	1, 3	1, 4	1, 5	1, 6
2	2, 1	**2, 2**	2, 3	2, 4	2, 5	2, 6
3	3, 1	3, 2	**3, 3**	3, 4	3, 5	3, 6
4	4, 1	4, 2	4, 3	**4, 4**	4, 5	4, 6
5	5, 1	5, 2	5, 3	5, 4	**5, 5**	5, 6
6	6, 1	6, 2	6, 3	6, 4	6, 5	**6, 6**

Six of the combinations are doubles.

The probability of throwing doubles is 6 / 36 = 1 /6.

The dice rolls are independent.

The probability of throwing three doubles in a row is $(1/6)^3$ = 1 / 216.

Data Analysis, Statistics, and Probability; Probability

(D) $1/108$

(E) $1/36$

20. What is the range of the function defined by

$$f(x) = \begin{cases} x^3, & x > -3 \\ 3x - 2, & x \leq 3 \end{cases}$$

(A) $-27 < y \leq -11$

(B) $y \leq -11$

(C) $y > -27$

(D) $y \leq -11$

*(E)   All real numbers.

To determine the range of a piecewise function, consider the range of all parts of the function.

The range of $f(x) = x^3$ for $x > -3$ is $y > -27$

The range of $f(x) = 3x - 2$ for $x \geq -3$ is $y \leq -11$

Combining $y > -27$ and $y \leq -11$ gives all real numbers as the range.

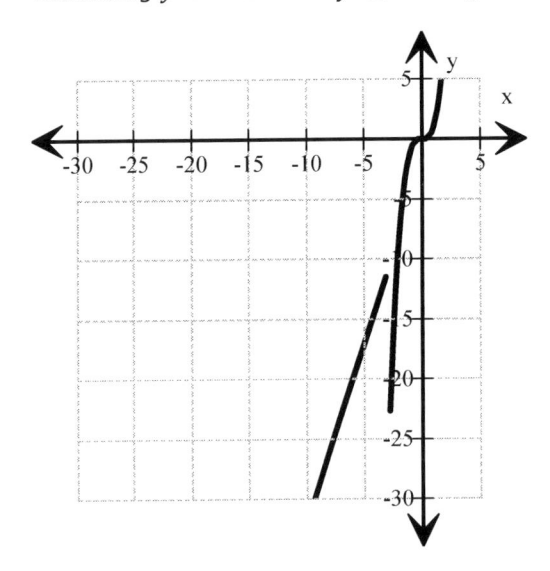

Algebra and Functions; Properties of Functions, Piecewise

21. Which of the following expressions is equivalent to $(\tan(2\alpha))(\tan^2 \alpha)$?

    (A) $\tan(2\alpha) - \tan \alpha$

    (B) $\tan \alpha + 2 \tan \alpha$

    (C) $\tan \alpha - 2 \tan \alpha$

    (D) $\tan(2\alpha) + 2 \tan \alpha$

    *(E) $\tan(2\alpha) - 2 \tan \alpha$

Double angle formulas:

$$\sin(2\alpha) = 2 \sin \alpha \cos \alpha$$

$$\cos(2\alpha) = \cos^2 \alpha - \sin^2 \alpha$$

$$\tan(2\alpha) = 2 \tan \alpha / (1 - \tan^2 \alpha)$$

Derivation of double angle formula for tangent:

$$\tan(2\alpha) = \sin(2\alpha) / \cos(2\alpha)$$

$$= (2 \sin \alpha \cos \alpha) / (\cos^2 \alpha - \sin^2 \alpha)$$

$$= (2 \sin \alpha / \cos \alpha) / (1 - \sin^2 \alpha / \cos^2 \alpha)$$

$$= 2 \tan \alpha / (1 - \tan^2 \alpha)$$

$(\tan(2\alpha))(1 - \tan^2 \alpha) = 2 \tan \alpha$

$(\tan(2\alpha)) - ((\tan 2\alpha)(\tan^2 \alpha)) = 2 \tan \alpha$

$\tan(2\alpha) = 2 \tan \alpha + ((\tan(2\alpha))(\tan^2 \alpha))$

$(\tan(2\alpha))(\tan^2 \alpha) = \tan(2\alpha) - 2 \tan \alpha$

$(\tan(2(\frac{\pi}{8})))(\tan^2(\frac{\pi}{8})) = \tan(2(\frac{\pi}{8})) - 2 \tan(\frac{\pi}{8})$

$(1)(0.4142)^2 = 0.1716; -2(0.4142) = 0.1716$

Geometry and Measurement; Trigonometry, Double Angle Formulas

22. A sequence is recursively defined by $a_n = a_{n-1} - 2a_{n-2}$, for $n > 2$. If $a_1 = 0$ and $a_2 = 1$, what is the value of a_{11}?

(A) -17

*(B) -11

n	1	2	3	4	5	6	7	8	9	10	11
$a_{n-1} - 2a_{n-2}$	0	1	1	-1	-3	-1	5	7	-3	-17	-11

Algebra and Functions; Properties of Functions, Recursive

(C) -3

(D) 5

(E) 7

23. If, in a right triangle, $\sin \theta = x/4$, where $0 < \theta < \pi/2$ and $0 < x < 4$, then $\cos \theta =$

*(A) $\dfrac{(\sqrt{16 - x^2})}{4}$

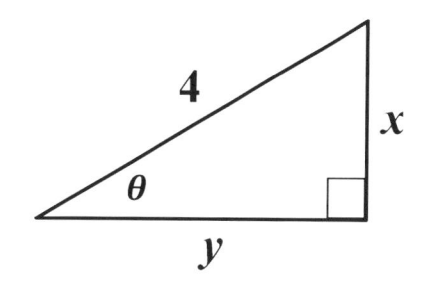

$x^2 + y^2 = 4^2 = 16$

$y = \sqrt{16 - x^2}$

$\cos \theta = \dfrac{(\sqrt{16 - x^2})}{4}$

$\sin^2 \theta + \cos^2 \theta = 1$

$\cos^2 \theta = 1 - \sin^2 \theta = 1 - (x/4)^2 = \dfrac{(16 - x^2)}{16}$

$\cos \theta = \dfrac{(\sqrt{16 - x^2})}{4}$

Geometry and Measurement; Trigonometry, Right Triangles

(B) $\dfrac{(\sqrt{16 + x^2})}{4}$

(C) $\dfrac{(\sqrt{4 - x^2})}{16}$

(D) $\dfrac{(\sqrt{4 + x^2})}{16}$

(E) $\dfrac{(\sqrt{4 - x^2})}{8}$

24. What is the period of the graph of $= 3 \tan(2\pi x - 5)$?

(A) $1/(3\pi)$

(B) $1/(2\pi)$

(C) $1/5$

*(D) $1/2$

The general form of the tangent function is:

$y = A \tan(Bx - C)$

Period T = the distance between any two repeating points on the function.

The period of the tangent function is: $T = \pi / |B|$

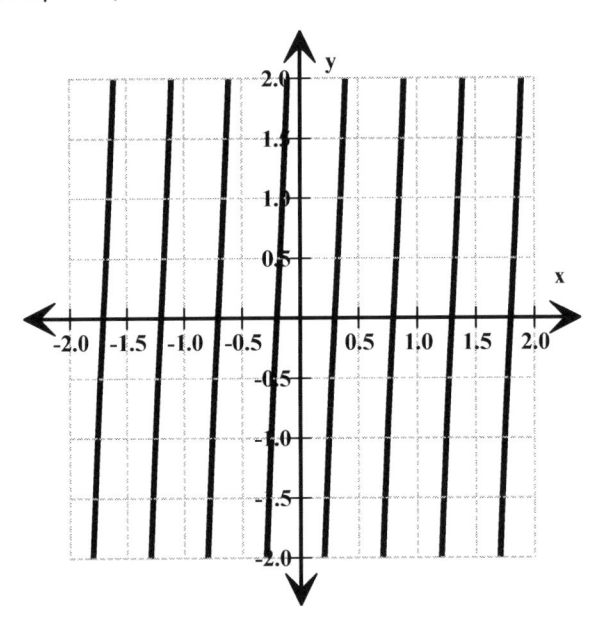

$T = \pi / |2\pi| = 1/2$

Algebra and Functions; Properties of Functions, Periodic

(E) $\pi/2$

25.

$$19\%, 8\%, -10\%, 14\%, -12\%, 0\%, 9\%, 25\%, 16\%, -1\%$$

If the percentiles are in the range 0 to 100, exclusive, what is the interquartile range for the rates of return shown above?

(A)   16%

(B)   17%

*(C)   20%

Excluding the percentiles at 0 and 100, the 25th percentile out of n terms is a weighted average of terms to get the term in $[(n + 1) / 4]^{th}$ place.

Order	1	2	3	4	5	6	7	8	9	10
Return	-12%	-10%	-1%	0%	8%	9%	14%	16%	19%	25%
Percentile	9	18	27	36	45	55	64	73	82	91

For $n = 10$, $[(n + 1) / 4] = (10 + 1) / 4 = 11 / 4 = 2.75$

The term in place 2.75 = (0.25) (−10%) + (0.75) (−1%) = −3.25%

Excluding the percentiles at 0 and 100, the 75th percentile out of n terms is a weighted average of terms to get the term in $[(3) (n + 1) / 4]^{th}$ place.

For $n = 10$, $[(3) (n + 1) / 4] = (3) (10 + 1) / 4 = 33 / 4 = 8.25$

The term in place 8.25 = (0.75) (16%) + (0.25) (19%) = 16.75%

The interquartile range is the absolute value of the difference between the 25th percentile and the 75th percentile.

16.75% − (−3.25%) = 20.00%

Data Analysis, Statistics, and Probability; Interquartile Range

(D)   29%

(E)   37%

26. If $f(x) = \sqrt[3]{2x^3 - 1}$, what is $f^{-1}(0.4)$?

(A) 0.65

*(B) 0.81

$$f^{-1}(0.4) = [((0.4)^3 + 1)/2]^{(1/3)} = 0.81$$

$$f(0.81) = \sqrt[3]{2(0.81)^3 - 1} = \sqrt[3]{0.062882} = 0.4$$

Algebra and Functions; Properties of Functions, Exponential

(C) 0.91

(D) 0.95

(E) 1.02

27.

The cube above is inscribed in a sphere. The cube has a surface area of 50 square inches. What is the volume of the sphere?

(A) 26.2

(B) 35.6

(C) 36.8

(D) 37.9

*(E) 65.4

Volume of a sphere with radius r: $V = \frac{4}{3} \pi r^3$

Surface area of a cube = $2lw + 2lh + 2wh$

$s = l = w = h$

$50 = = 2lw + 2lh + 2wh = 6s^2$

$s = (50/6)^{(1/2)} = 5/\sqrt{3} \approx 2.88675$

The diagonal of the cube is the hypotenuse of a triangle formed by an edge and the diagonal of a side.

The diagonal of a side = $s\sqrt{2}$

The diagonal of the cube = $s\sqrt{3} = 5$

The radius of the sphere is equal to 1/2 the diagonal of the cube = 5/2.

$V = \frac{4}{3} \pi r^3 = \frac{4}{3} \pi (\frac{5}{2})^3 = 65.4$

Geometry and Measurement; Surface Area and Volume of Solids, Spheres

28. As of September 2005, a company's revenue was $6.4 million. Assuming a growth rate of 5% per year, the company's revenue, in millions, for n years after 2010 can be modeled by the equation $R = \$6.4 \,(1.05)^n$. According to the model, what was the company's revenue growth from September 2009 to September 2010?

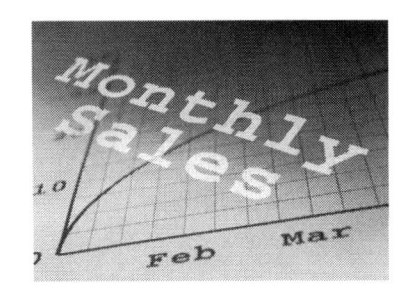

*(A) $389,000

($6,400,000) $[(1.05)^5 - (1.05)^4]$ = ($6,400,000) (0.060775)

= $388,962 ≈ $389,000

Revenue as of September 2010 = ($6,400,000) $(1.05)^5$ = $8,168,202

Revenue as of September 2009 = ($6,400,000) $(1.05)^4$ = $7,779,240

Revenue growth from September 2009 to September 2010 =

$8,168,202 − $7,779,240

= $388,962 ≈ $389,000

Algebra and Functions; Properties of Functions, Exponential

(B) $397,000

(C) $408,000

(D) $417,000

(E) $429,000

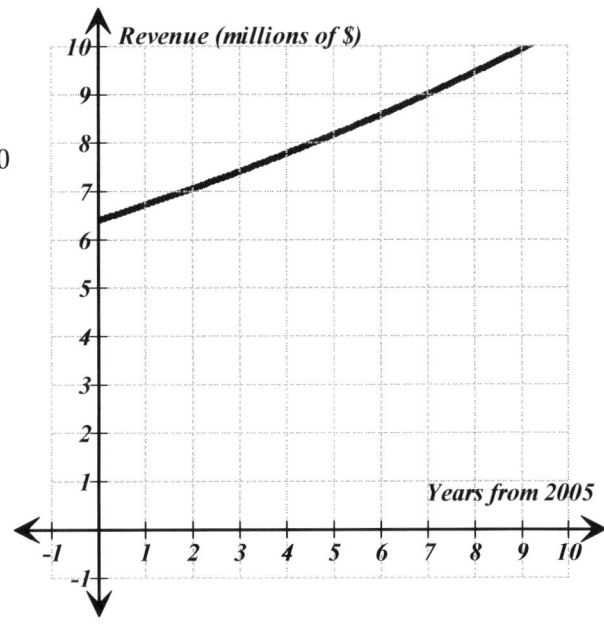

29. What is the measure of one of the larger angles of a parallelogram in the xy-plane that has vertices with coordinates (6, −3), (3, −7), (7, 7), and (4, −3) ?

(A)　　106.0°

(B)　　106.2°

*(C)　　106.7°

The smaller angle of the 90° triangle with vertices with coordinates (6, −3), (3, −7), and (−3, −3) has a tangent equal to 3 / 10.

$$arctan\left(\frac{3}{10}\right) = 16.7°$$

The smaller angle of the triangle plus 90° equals the measure of one of the larger angles of the parallelogram.

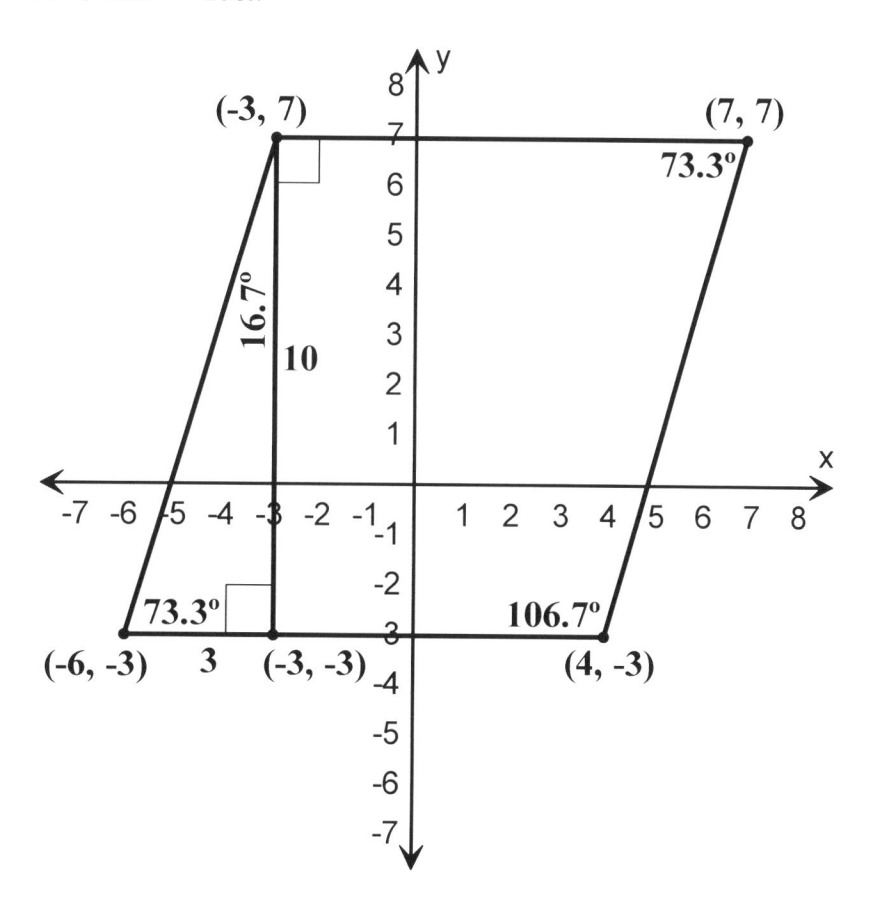

$$90° + 16.7° = 106.7°$$

Apply the Law of Cosines to the triangle formed by (6, −3), (3, −7), and (7, 7).

Law of Cosines: $c^2 = a^2 + b^2 - 2ab(\cos C)$

$c^2 = (7-(-3))^2 + (7-(-6))^2 = 10^2 + 13^2 = 269$

$a^2 = (7-(-3))^2 + (-3-(-6))^2 = 10^2 + 3^2 = 109$

$b = 7 - (-3) = 10 \;;\; b^2 = 10^2 = 100$

$269 = 109 + 100 - 2(\sqrt{109})(10)(\cos C) \; 209 - 208.8(\cos C)$

$\cos C = 60/208.8$

$arccos\left(\dfrac{60}{208.8}\right) = C = 73.3°$

$180° - 73.3° = 106.7°$

Geometry and Measurement; Trigonometry, Law of Cosines

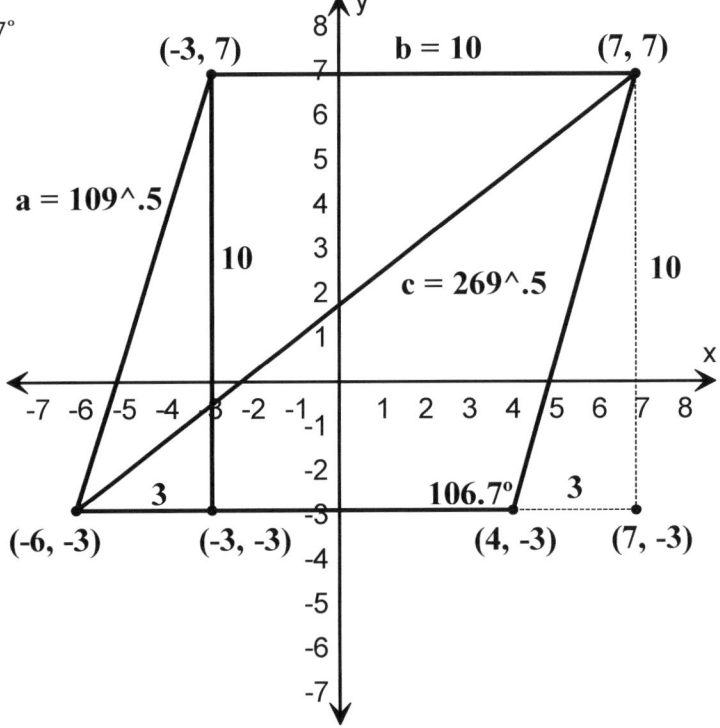

(D)   106.9°

(E)   108.0°

30. The population of the United States is projected to grow exponentially to 400 million in the year 2050, a 30% increase from the population in the year 2010. What is the projected population in the year 2040, in millions?

(A) 368

(B) 370

(C) 374

*(D) 375

$(1+g)^{40} = 1.3$

$40 \ln(1+g) = \ln 1.3$

$\ln(1+g) = \dfrac{\ln 1.3}{40}$

≈ 0.006559107

$e^{0.006559107} \approx 1.0065807$

$= 1 + g$

$g = 0.0065807 = 0.65807\%$

$(400/1.3)(1.0065807)^{30} = \left(\dfrac{400}{1.3}\right)(1.2175) \approx 375$

Algebra and Functions; Properties of Functions, Exponential

(E) 377

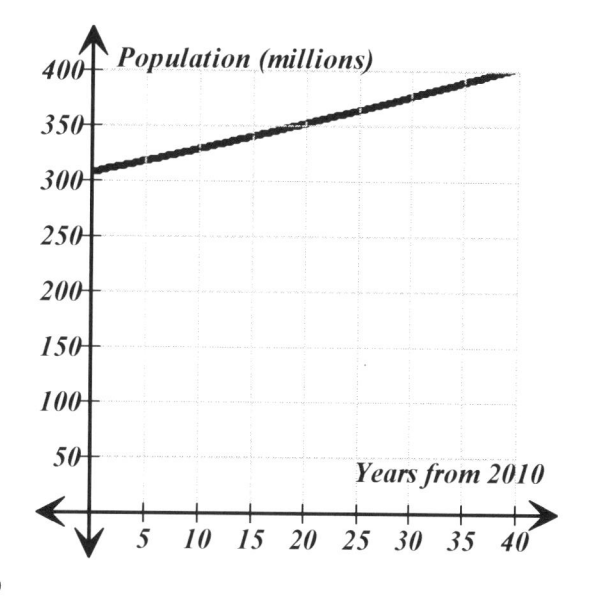

31. Find the exponential function that best fits the data below.

x	0	1	2	3	4	5
y	1.7	0.5	0.16	0.05	0.016	0.005

*(A) $y = (1.65)(0.31^x)$

Using the TI-83 Plus calculator:

STAT

ENTER

$L1(1) = 0, L1(2) = 1, L1(3) = 2, L1(4) = 1, L1(5) = 4, L1(6) = 5$

$L2(1) = 1.7, L2(2) = 0.5, L2(3) = 0.16, L2(4) = 0.05, L2(5) = 0.016, L2(6) = 0.005$

STAT

CALC

0

ExpReg

ENTER

The screen should show the following:

ExpReg
     y = a * b ^ x
     a = 1.646038198
     b = .3131823628

$y = (1.65)(0.31^x)$

Data Analysis, Statistics, and Probability; Least Squares Regression, Exponential

(B) $y = (1.68)(0.31^x)$

(C) $y = (1.65)(0.32^x)$

(D) $y = (1.68)(0.32^x)$

(E) $y = (1.70)(0.30^x)$

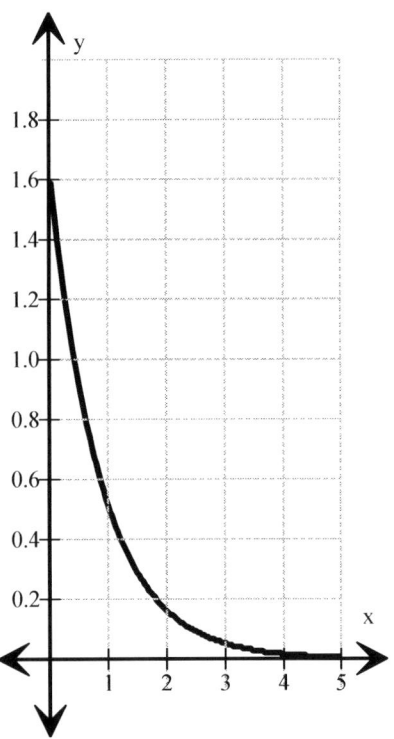

32. After eliminating the parameter, what is the rectangular equation whose graph represents the parametric curve $x = t^2/2$ and $= t^2 + 1$?

(A) $y = 4x^2 + 1$ for all real numbers

(B) $y = 4x^2 + 1$ for $x \geq 0$

(C) $y = 2x + 1$ for all real numbers

*(D) $y = 2x + 1$ for $x \geq 0$

t	−3	−2	−1	0	1	2	3
x	4.5	2.0	0.5	0.0	0.5	2.0	4.5
y	10	5	2	0	2	5	10

$x = t^2/2$

$2x = t^2$

$y = t^2 + 1 = 2x + 1$ for $x \geq 0$

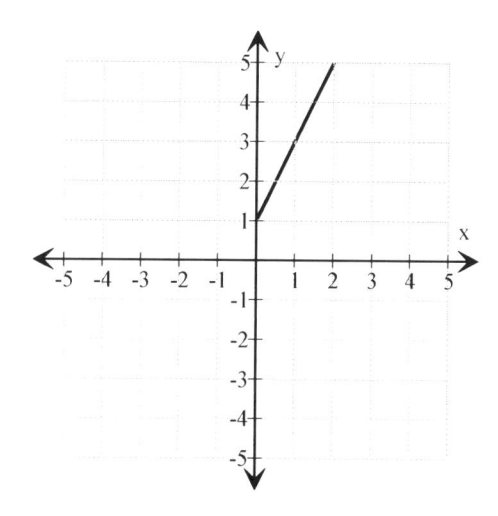

Algebra and Functions; Properties of Functions, Parametric

(E) $y = (x - 1)^2/2$ for $x \geq 1$

33.

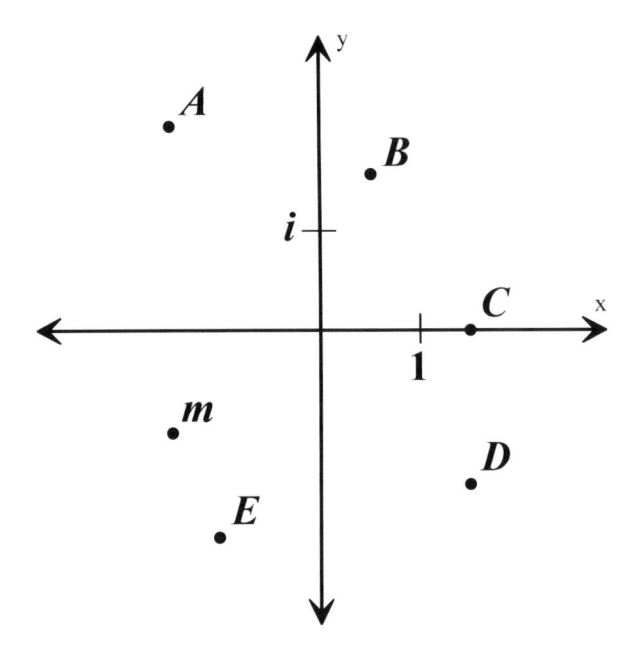

If m is the complex number shown in the figure above, which of the following could be im?

(A) A

(B) B

(C) C

*(D) D

The complex number given by m is equal to $a + bi$, where $a < 0$ and $b < 0$.

Multiplying by i will give $ai + bi^2 = ai - b$.

The x-coordinate of im equals $-b$, and $-b > 0$.

The y-coordinate of im equals ai, and $a < 0$.

Number and Operations; Complex Numbers

(E) E

34. Which of the following graphs represents the greatest integer function, $y = [x]$, where $[x]$ is the greatest integer less than or equal to x ?

(A)

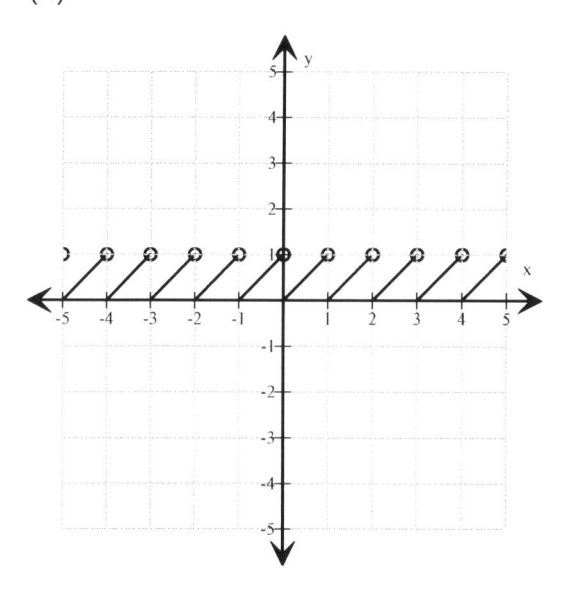

(B)

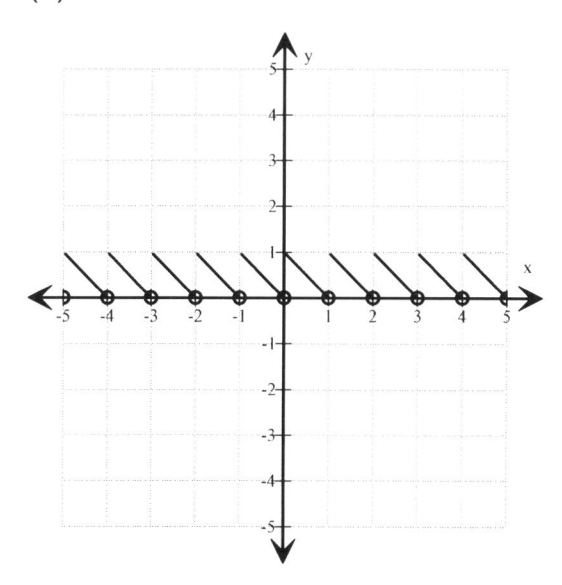

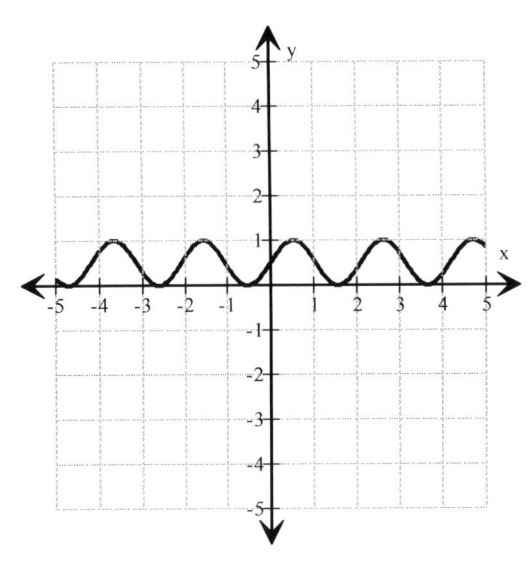

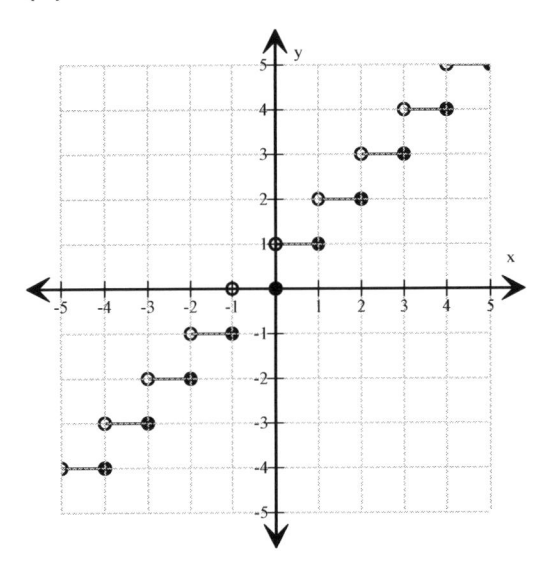

(C)

(D)

*(E)

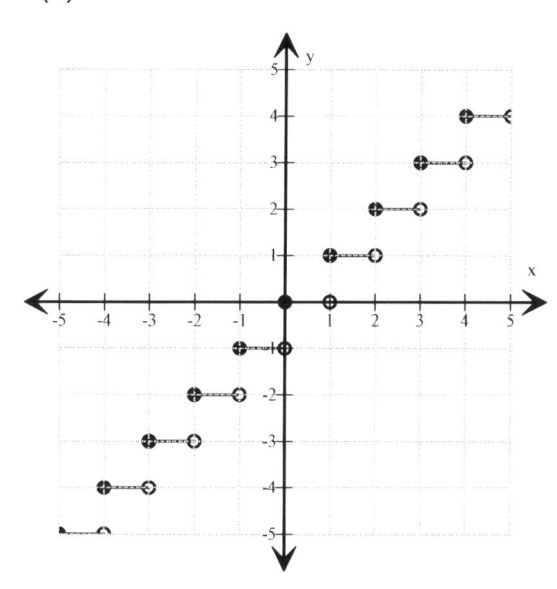

Graph A is the mantissa function $y = x - [x]$

Graph B is a piecewise function with a period of 1 given by $y = -x + n$ for the domain $n - 1 \leq x < n$, where n is an integer.

Graph C is a trigonometric function.

Graph D is a piecewise function where y is the smallest integer not less than x.

Graph E is the piecewise function known as the floor function or greatest integer function, $y = [x]$

If the number is an integer, the function equals that integer.

If the function is not an integer, the function equals the next lowest integer.

Algebra and Functions; Properties of Functions, Piecewise

35. Express the point with polar coordinates $(2, \frac{3\pi}{10})$ in terms of rectangular coordinates.

(A)   (0.6, 0.8)

(B)   (1.0, 1.7)

*(C)   (1.2, 1.6)

$x = r \cos \theta$

$y = r \sin \theta$

$x^2 + y^2 = r^2$

$x = 2 \cos \left(\frac{3\pi}{10} \right)$

$= 2(0.5878) \approx 1.2$

$y = 2 \sin \left(\frac{3\pi}{10} \right)$

$= 2(0.80890) \approx 1.6$

$(x, y) = (1.2, 1.6)$

$1.2^2 + 1.6^2 = 2.0^2$

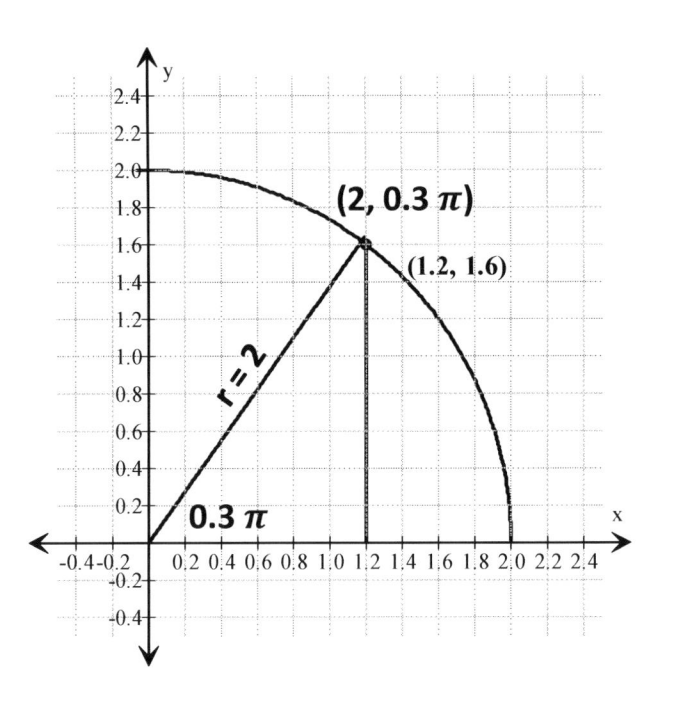

$\theta = 0.3\pi = 54°$

Geometry and Measurement; Coordinate, Polar Coordinates

(D)   (1.4, 1.6)

(E)   (1.6, 1.2)

36. The population of a colony of bacteria increases exponentially. There are 7,000 bacteria at the start. The formula $N = N_0 \, e^{0.06\,t}$ gives the population after t hours. To the nearest hour, how many hours will it take for the population to reach 10,000?

(A) 5

*(B) 6

$$10{,}000 = 7{,}000 \, e^{0.06\,t}$$

$$e^{0.06\,t} = \frac{10{,}000}{7{,}000} = \frac{10}{7}$$

$$\ln e^{0.06\,t} = \ln \frac{10}{7}$$

$$0.06\,t = \ln \frac{10}{7} = 0.356675$$

$$t = \frac{0.356675}{0.06} = 5.94 \approx 6$$

$$7{,}000 \, e^{((0.06)(6))} = 10{,}033$$

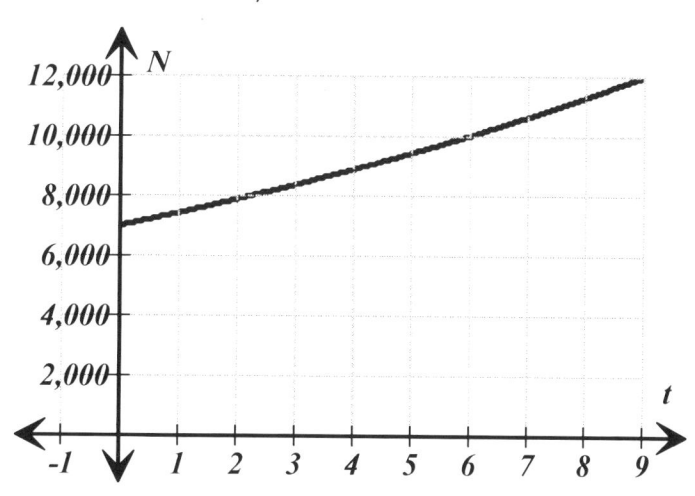

Algebra and Functions; Properties of Functions, Logarithmic

(C) 7

(D) 8

(E) 9

37. A lottery pays $1 per year for an infinite number of years. What is the present value of the payment stream if the cash flows are discounted at an interest rate of 8% per year and the first of the annual $1 payments starts immediately?

(A)　　$8.64

(B)　　$10.00

(C)　　$12.50

*(D)　　$13.50

($1 / 0.08) + $1.00 = $12.50 + $1.00 = $13.50

The present value is the sum of an infinite geometric series.

$$S_n = \frac{\$1}{1.08^0} + \frac{\$1}{1.08^1} + \frac{\$1}{1.08^2} + \frac{\$1}{1.08^3} + \cdots = \frac{a_1}{1-r}$$

$a_1 = \$1$

$r = \dfrac{1}{1.08}$

$$S_n = \frac{a_1}{1-r} = \frac{\$1}{1-\left(\frac{1}{1.08}\right)} = \frac{\$1}{\left(\frac{1.08-1}{1.08}\right)} = \frac{\$1}{\left(\frac{0.08}{1.08}\right)}$$

$$S_n = \frac{\$1\,(1.08)}{0.08} = \$12.50\,(1.08) = \$13.50$$

Number and Operations; Series

(E)　　$25.00

38. If $0 < \theta < \pi/2$ and $\cos\theta = x$, then $\tan\theta =$

(A) $x/(\sqrt{1-x^2})$

(B) $x/(1-x^2)$

(C) $1/x$

*(D) $(\sqrt{1-x^2})/x$

$$\sin^2\theta + \cos^2\theta = 1$$

$$\sin^2\theta = 1 - \cos^2\theta = 1 - x^2$$

$$\sin\theta = \sqrt{1-x^2}$$

$$\tan\theta = \frac{\sin\theta}{\cos\theta} = (\sqrt{1-x^2})/x$$

$$\cos\left(\frac{\pi}{4}\right) = 0.7071$$

$$\tan\left(\frac{\pi}{4}\right) = 1.0000$$

$$\frac{\sqrt{1-(0.7071)^2}}{0.7071} = 1$$

Geometry and Measurement; Trigonometry, Equations

(E) $(1-x^2)/x$

39. Which of the following graphs represents the curve determined by the parametric equations $x = 2t - 3$ and $y = t^2 - 5$ on the interval $0 \leq t \leq 3$?

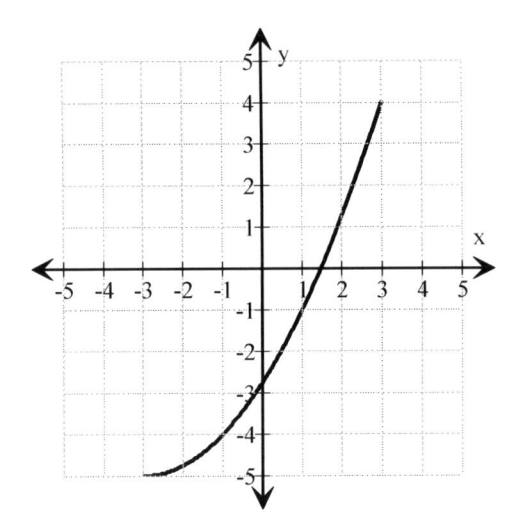

*(A)

t	0	1	2	3
x	−3	−1	1	3
y	−5	−4	−1	4

$x = 2t - 3$

$x + 3 = 2t$

$t = \dfrac{(x+3)}{2}$

$y = t^2 - 5 = [\dfrac{(x+3)}{2}]^2 - 5$ for $-3 \leq x \leq 3$

Algebra and Functions; Properties of Functions, Parametric

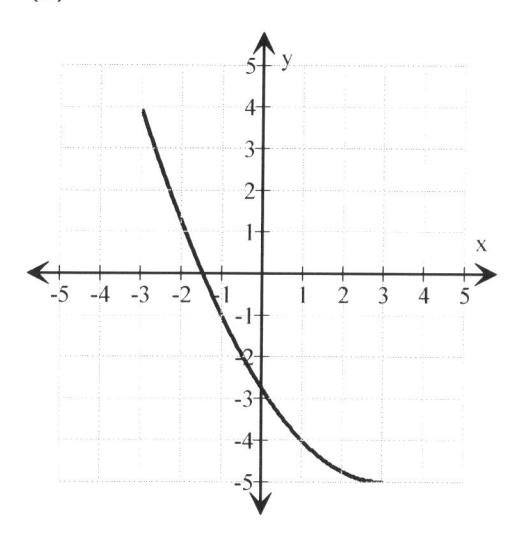

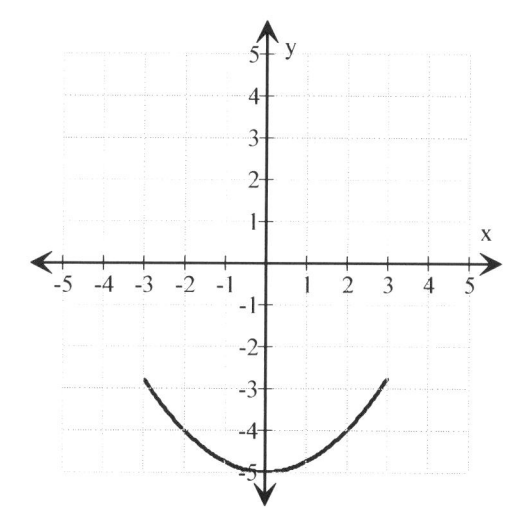

(B)

(C)

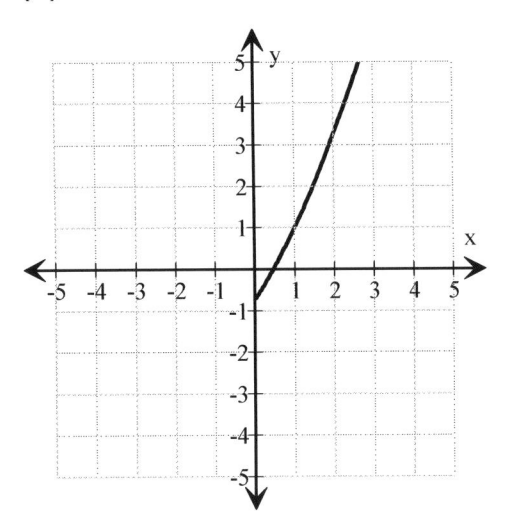

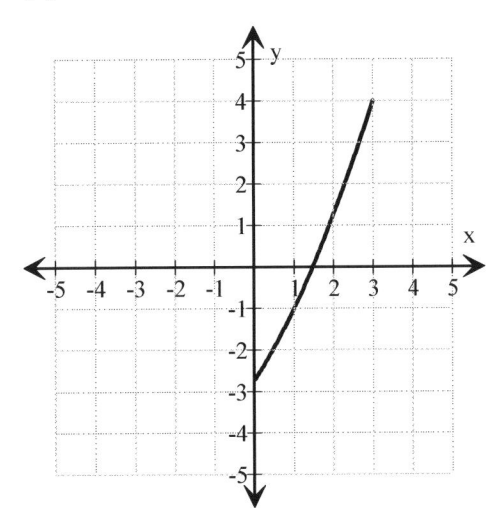

(D)

(E)

40. Using the quadratic regression model, which parabola best fits the data for the number of Hummers sold in the U.S., in thousands, and the number of years since 2000?

Years since 2000	0	1	2	3	4	5	6	7	8	9	10
Number Sold (thousands)	1	1	20	35	29	57	72	56	27	8	2

(A) $y = -3x^2 + 26x - 15$

*(B) $y = -2x^2 + 24x - 13$

Using the TI-83 Plus calculator:

STAT
ENTER

L1(1) = 0, L1(2) = 1, L1(3) = 2, L1(4) = 1, L1(5) = 4, L1(6) = 5, L1(7) = 6, L1(8) = 7, L1(9) = 8, L1(10) = 9, L1(11) = 10

L2(1) = 1, L2(2) = 1, L2(3) = 20, L2(4) = 35, L2(5) = 29, L2(6) = 57, L2(7) = 72, L2(8) = 56, L2(9) = 27, L2(10) = 8, L2(11) = 2

STAT
CALC
5
QuadReg
ENTER

The screen should show the following:

ExpReg
$y = ax^2 + bx + c$
a = -2.2995338
b = 24.25897436
c = -12.81118881

$y = -2x^2 + 24x - 13$

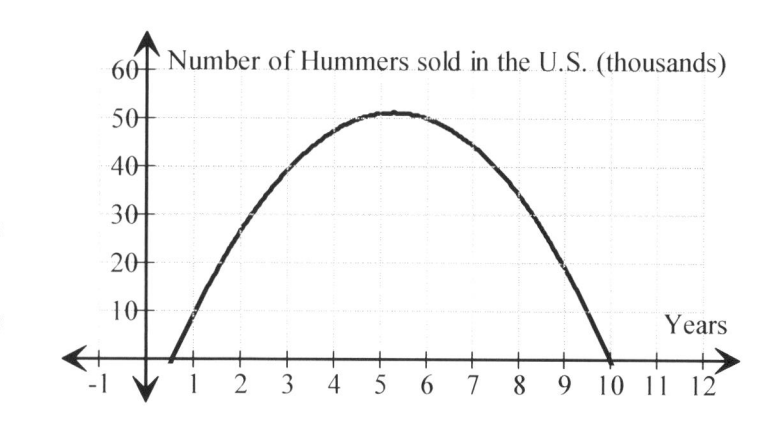

Data Analysis, Statistics, and Probability; Least Squares Regression, Quadratic

(C) $y = 2x^2 - 26x - 13$

(D) $y = 2x^2 - 24x + 13$

(E) $y = 3x^2 - 26x + 15$

41. What is the length of the major axis of the ellipse whose equation is
$48x^2 + 16y^2 = 112$

(A) 1.7

(B) 2.6

(C) 3.5

*(D) 5.3

The general equation for an ellipse centered at the origin is:

$ax^2 + by^2 = c$ where $a \ne b$ and a, b, and c are all positive.

$a = 48$, $b = 16$, $c = 112$

$y^2 = (112 - 48x^2) / 16$

$= 7 - 3x^2$

$y = (7 - 3x^2)^{0.5}$

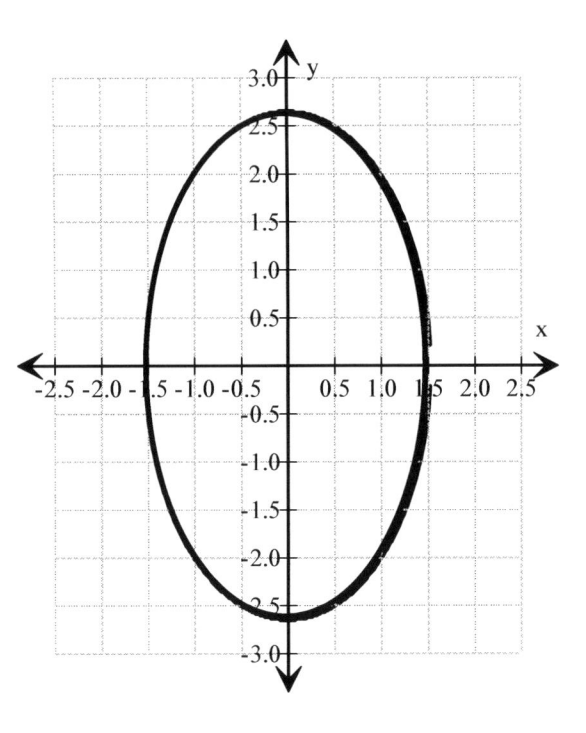

The major axis is vertical with vertices $(0, \sqrt{7})$ and $(0, -\sqrt{7})$.

The length of the major axis = $\sqrt{7} - (-\sqrt{7}) = 2\sqrt{7} \approx 5.3$

When $= 0$, $16(y^2) = 112$

$y^2 = \dfrac{112}{16} = 7$

$y = \sqrt{7}$

$2y = 2\sqrt{7} \approx 5.3$

Geometry and Measurement; Coordinate, Ellipses

(E) 6.9

42. In a savings account that earns 4% compounded annually, how many years would it take for a customer to triple the amount initially deposited?

(A) 18

(B) 21

*(C) 28

$1.04^n = 3$

$n \ln 1.04 = \ln 3$

$n = \dfrac{\ln 3}{\ln 1.04} \approx 28$

$1.04^{28} = 3$

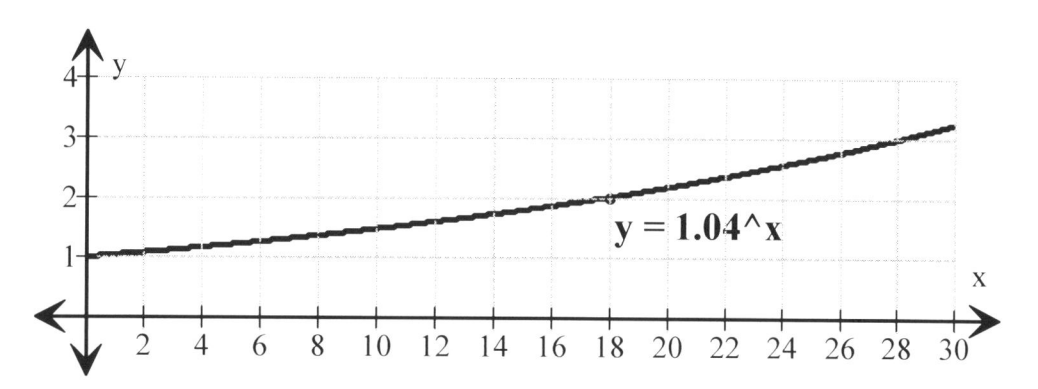

Algebra and Functions; Properties of Functions, Exponential

(D) 32

(E) 36

43. A standard 52-card deck of playing cards has four aces. What is the probability of being dealt a five-card hand that includes four aces?

(A) $1/270{,}725$

*(B) $1/54{,}145$

The probability of being dealt four aces in a five-card hand is equal to the probability of being dealt four aces in a four-card hand, times five. The card that is not an ace can be dealt first, second, third, fourth, or fifth.

$$(5)\left(\tfrac{4}{52}\right)\left(\tfrac{3}{51}\right)\left(\tfrac{2}{50}\right)\left(\tfrac{1}{49}\right) = \left(\tfrac{120}{6{,}497{,}400}\right) = \left(\tfrac{1}{54{,}145}\right)$$

A combination is a way of selecting several things from a group where order does not matter.

$$\binom{n}{k} = \frac{(n)(n-1)\ldots(n-k+1)}{(k)(k-1)\ldots(1)} = \frac{n!}{k!\,(n-k)!}$$

The factorial for a non-negative number n is denoted by $n!$ and is defined as the product of all positive integers less than or equal to n.

$$n! = (n)(n-1)\ldots(1)$$

The distinct number of five-card hands $= 52!/(5!\,47!)$

$$= (52)(51)(50)(49)(48)/(5)(4)(3)(2) = 2{,}598{,}960$$

For a hand with four aces, the fifth card could be one of the 48 cards that is not an ace. The number of hands with four aces = 48

The probability of being dealt four aces in a five card hand

$$= 48/2{,}598{,}960 = 1/54{,}145$$

Data Analysis, Statistics, and Probability; Probability

(C) $1/20{,}825$

(D) $1/16{,}660$

(E) $1/4{,}165$

44. Which of the following graphs is the best parametric representation of the equations $x = 3\sec(t)$ and $y = \tan(t)$?

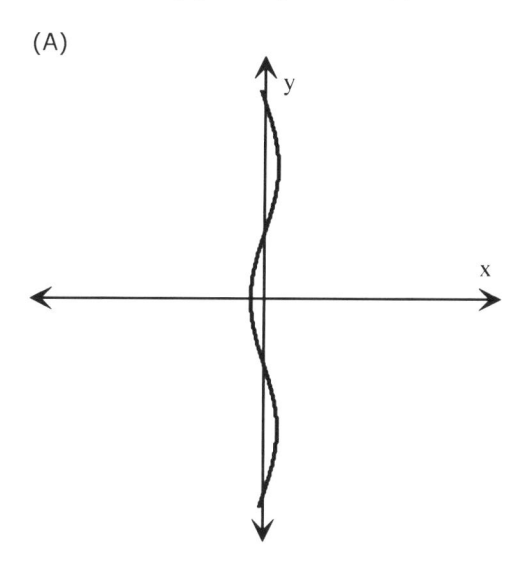

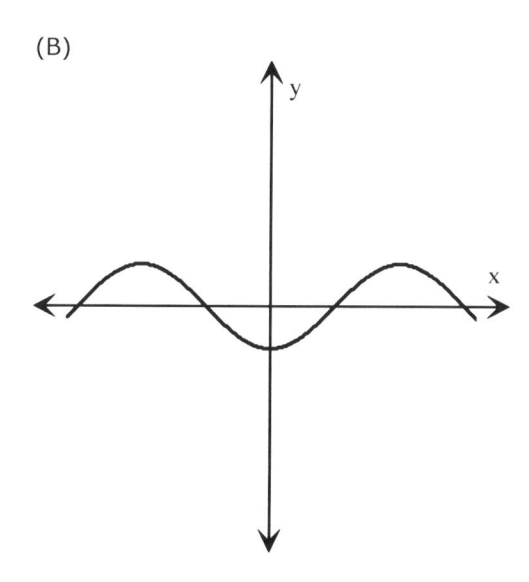

(A)

(B)

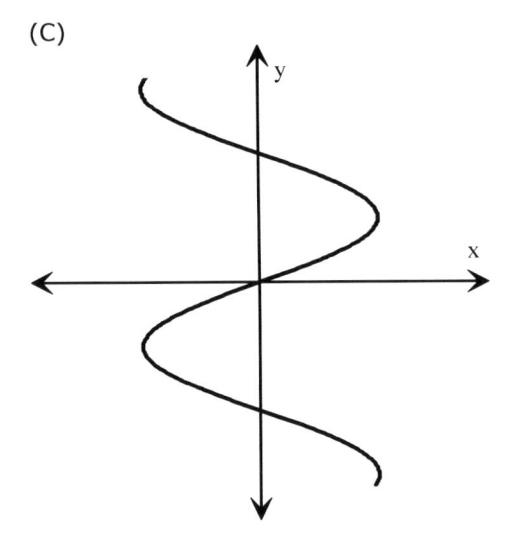

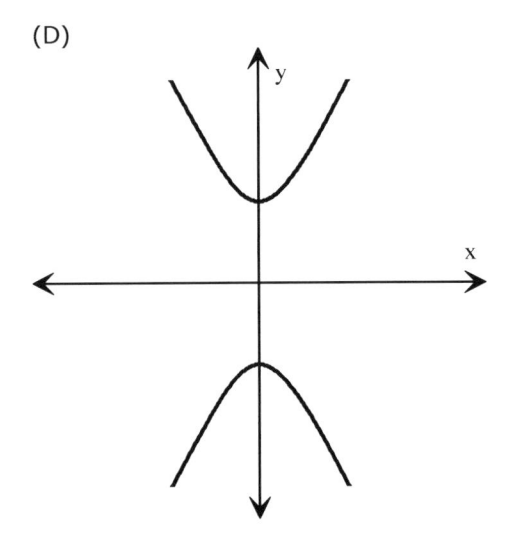

(C)

(D)

*(E)

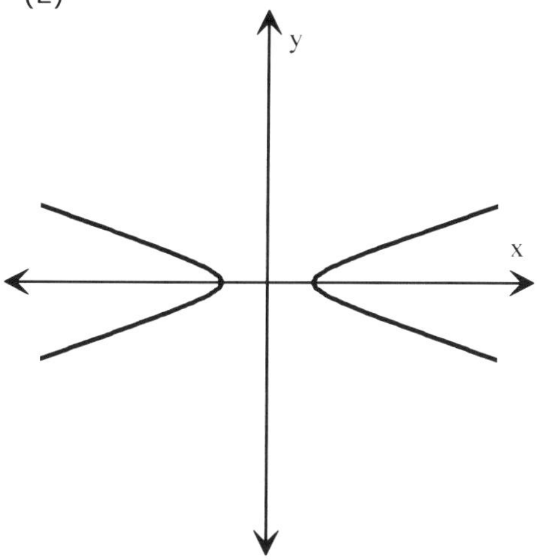

$$\sec^2(t) = 1 + \tan^2(t)$$

$\sec^2(t) - \tan^2(t) = 1$

$y = tan\,(t)$

$y^2 = \tan^2(t)$

$x = 3\,sec\,(t)$

$x^2 = 9\sec^2(t)$

$\sec^2(t) = \dfrac{x^2}{9}$

$\dfrac{x^2}{9} - y^2 = 1$

The equations are the parametric representation of a left and right hyperbola.

Algebra and Functions; Properties of Functions, Parametric

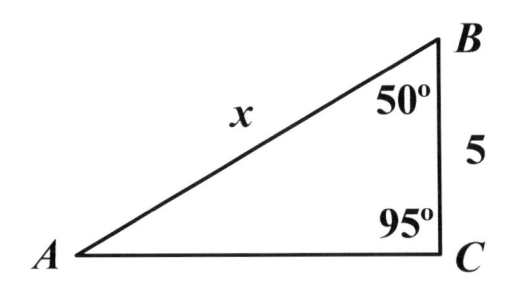

45.

In the figure above, $BC = 5$, $\angle B = 50°$, and $\angle C = 95°$. What is the value of x ?

(A) 6.53

(B) 6.68

*(C) 8.68

Law of Sines: $\sin A / BC = \sin B / AC = \sin C / AB$

$\angle A = 180° - 50° - 95° = 35°$

$\sin 35° / 5 = \sin 95° / x$

$x = 5 \sin 95° / \sin 35° = (5)(0.99619) / (0.57358) = 8.684$

Angle	A	B	C
Opposite Side	BC	AC	AB
Measure (degrees)	35	50	95
sin	0.57358	0.76604	0.99619
Opposite side length	5.000	6.678	8.684
sin / opposite side	0.1147	0.1147	0.1147

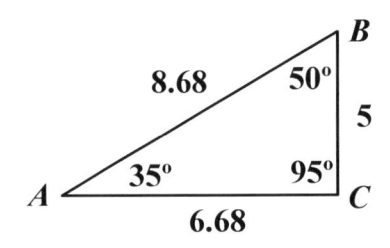

Triangle ABC can be split into 2 right triangles ACD and BCD.

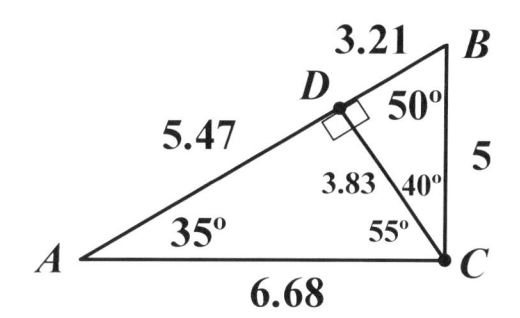

∠ BCD = 180° − 90° − 50° = 40°

sin 40° = DB / 5.00

0.64278761 = DB / 5.00

BD = (0.64278761) (5.00) ≈ 3.213938

$DC^2 + 3.213938^2 = 5^2$

$DC = \sqrt{25 - 10.329937} = 3.830222$

∠ ACD = 180° − 90° − 35° = 55°

tan 55° = AD / 3.830222

1.42814801 = AD / 3.830222

AD = (1.42814801) (3.830222) ≈ 5.470

AB = AD + DB = 5.470 + 3.214 = 8.684

Geometry and Measurement; Trigonometry, Law of Sines

(D)   8.72

(E)   8.75

46. Suppose the graph of $f(x) = x^2 - 2x + 2$ is translated 2 units right and 1 unit down. If the resulting graph represents $g(x)$, what is the value of $g(1.5)$?

*(A)   2.25

$$g(x) + 1 = (x - 2)^2 - 2(x - 2) + 2$$

$$g(1.5) = (1.5 - 2)^2 - 2(1.5 - 2) + 2 - 1 = 0.25 + 1 + 1 = 2.25$$

$$g(x) + 1 = x^2 - 4x + 4 - 2x + 4 + 2$$

$$g(x) = x^2 - 6x + 9$$

$$g(1.5) = (1.5)^2 - 6(1.5) + 9 = 2.25 - 9 + 9 = 2.25$$

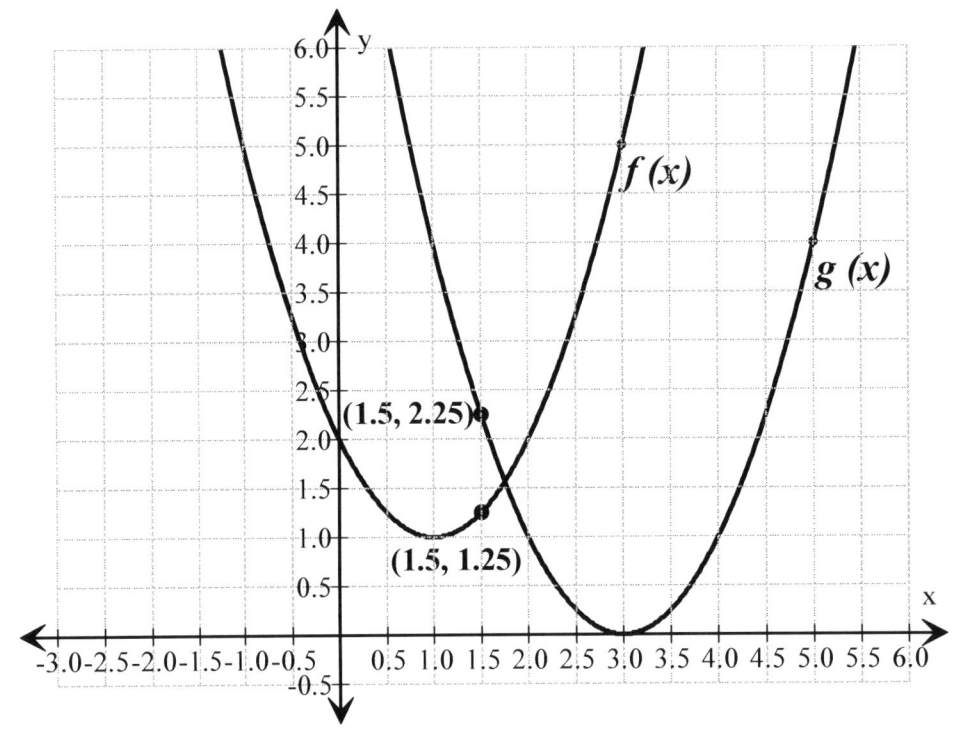

Geometry and Measurement; Coordinate, Transformations

(B)   3.25

(C)   4.25

(D)   6.25

(E)   8.25

MCCAULAY'S PRACTICE EXAMS FOR THE SAT* SUBJECT TEST IN MATHEMATICS LEVEL 2

47. Which of the following functions represents a sine function with a maximum of 4, a minimum of −2, a period of $\pi/4$, and no horizontal shift?

*(A) $f(x) = 3 \sin(8x) + 1$

The general form of the sine function is:

$y = A \sin(Bx + C) + D$

Amplitude A = the distance from the midpoint to the highest or lowest point of the function.

Period T = the distance between any two repeating points on the function.

$T = 2\pi / |B|$

$A = (4 - (-2))/2 = 3$

$T = \pi/4 = 2\pi / |B|$

$B > 0$ in all the choices; $B = 2\pi / (\pi/4) = 8$

D = Vertical displacement, or 'y' shift, of the midpoint of the function above the x-axis = $(4 + (-2))/2 = 1$.

Phase shift = the amount of horizontal displacement of the function from its original position = $C/B = 0$

$C = 0$

$f(x) = A \sin(Bx + C) + D$
$= 3 \sin(8x) + 1$

Algebra and Functions;
Properties of Functions,
Trigonometric

(B) $f(x) = 3 \sin(4x) - 1$

(C) $f(x) = 3 \sin(4x) + 2$

(D) $f(x) = 6 \sin(3x) - 1$

(E) $f(x) = 6 \sin(8x) + 1$

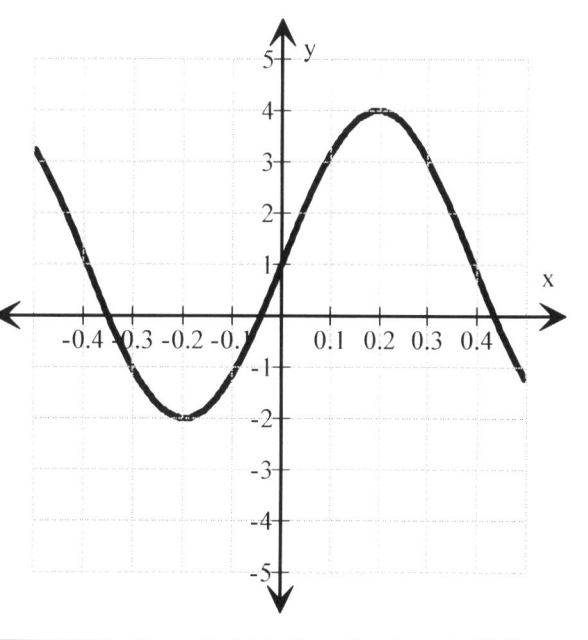

Page 181

48. A new employee starts a job with an annual salary of $100,000. The first pay increase will be granted after one year. Each annual pay increase is 5% of the prior year's salary. How much will the employee earn in total salaries over a 30 year career?

(A) $3,021,856

(B) $5,175,000

(C) $6,232,271

*(D) $6,643,885

The total over 30 years is the sum of a finite geometric series.

$S_n = \$100,000 + \$100,000(1.05) + \$100,000(1.05)^2 + \cdots + \$100,000(1.05)^{29}$

$a_1 = \$100,000$

$r = 1.05$

$S_n = \dfrac{a_1(1-r^n)}{1-r} = \dfrac{\$100,000\,(1-1.05^{30})}{1-1.05} = \dfrac{\$100,000\,(-3.32194)}{-0.05}$

$S_n = \$100,000\,(66.43885) = \$6,643,885$

Number and Operations; Series

(E) $6,976,079

49. The polyhedron shown below consists of a cube with sides of length S with an identical set of eight cubes attached to each of the six faces of the cube, for a total of 49 cubes. A total of 174 square pyramids with the lengths of the sides of the pyramid bases equal to the lengths of the sides of the cubes will be attached to 174 of the 198 exposed cube faces to make a 720-faced polyhedron. The height of each of the 174 square pyramids equals $(s^4 - 49s)/58$. What is the volume of the polyhedron formed by the 49 cubes and the 174 square pyramids?

(A) $78 s^3$

(B) $49 s^3 + 58 s^4$

(C) $49 s^3 + s^6$

(D) $49 s^3 + 174 s^2$

*(E) s^6

Volume of a pyramid with base area B and height h: $V = \frac{1}{3} B h$

$$B = s^2$$

$$V = 49\,s^3 + 174\left(\frac{1}{3}s^2 H\right) = 49\,s^3 + 58\,s^2 H$$

$$H = \frac{s^4 - 49\,s}{58}$$

$$V = 49\,s^3 + 58\,s^2 H$$

$$V = 49\,s^3 + 58\,s^2 \left[\frac{s^4 - 49\,s}{58}\right]$$

$$V = 49\,s^3 + s^2[s^4 - 49\,s]$$

$$V = 49\,s^3 + s^6 - 49\,s^3$$

$$V = s^6$$

Geometry and Measurement; Surface Area and Volume of Solids, Pyramids

50.

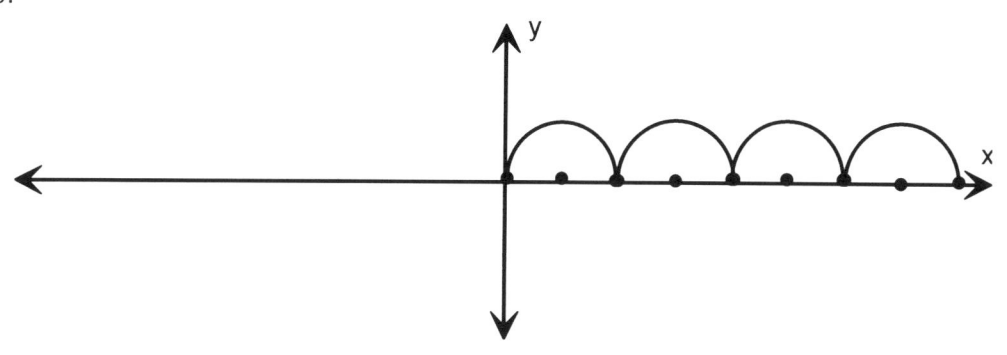

Which of the following parametric equations is represented by the above graph?

(A) $x = \alpha(t - \sin(t))$ and $y = \alpha(t - \cos(t))$

(B) $x = \alpha(1 + \sin(t))$ and $y = \alpha(1 + \cos(t))$

(C) $x = \alpha(t + \sin(t))$ and $y = \alpha(t + \cos(t))$

(D) $x = \alpha(1 - \sin(t))$ and $y = \alpha(t - \cos(t))$

*(E) $x = \alpha(t - \sin(t))$ and $y = \alpha(1 - \cos(t))$

$$\frac{y}{\alpha} = 1 - \cos(t)$$

$$\cos(t) = 1 - \frac{y}{\alpha}$$

$$t = \cos^{-1}\left(1 - \left(\frac{y}{\alpha}\right)\right)$$

$$\sin(t) = \sin\left(\cos^{-1}\left(1 - \left(\frac{y}{\alpha}\right)\right)\right)$$

The Cartesian equation for a cycloid is:

$$x = \alpha \cos^{-1}\left(1 - \left(\frac{y}{\alpha}\right)\right) - \sqrt{2\alpha y - y^2}$$

The cycloid is a point on a circle of radius α rolling along a straight line.

The equations are the parametric representation of a cycloid.

Algebra and Functions; Properties of Functions, Parametric

About the Author

Philip Martin McCaulay is an actuary with a degree in mathematics from Indiana University. He has sold thousands of study guides and practice exam books in the fields of math, pensions, investments, finance, real estate, and massage therapy. He has also published books on card games, cooking, and military history. He volunteers to write, publish, and ship free copies of books to troops and military families through Operation Paperback.

He is a Fellow of the Society of Actuaries, an Enrolled Actuary, a Member of the American Academy of Actuaries, and a Fellow of the Conference of Consulting Actuaries, with experience as a Vice Chair for the Society of Actuaries Education & Examination Committee.

Acknowledgements: *The author is grateful to Leslie Stewart, Michael Chambers, and Phillip Niles McCaulay for proofreading the book; to Anna Shadinger for being the model; and to Leslie Stewart for the photography.*

Printed in Great Britain
by Amazon